21世纪高等学校规划教材 | 计算机应用

Android移动应用开发教程

祝永志 主编
申 健 朱盼盼 刘梦芸 副主编

清华大学出版社
北京

内容简介

本书详细阐述了基于Android操作系统的移动应用开发技术，共9章。第1章主要介绍Android的基础知识；第2章讲解Android开发环境的搭建以及不同环境之间的转换与比较等；第3章讲述Activity及其生命周期，JUnit单元测试，资源的调用等；第4章讲解常见的UI控件以及自定义控件的使用等；第5章讲解Intent与组件通信；第6章讲解Android的后台服务；第7章讲解数据存储技术；第8章讲解网络通信技术，包括Android网络通信原理，Socket、HTTP、URL以及WebView等网络通信机制等；第9章是一个完整的综合案例——移动办公软件系统。

本书适合Android开发的初学者，尤其适合作为高等院校计算机本、专科及相关专业的教材，也可供有一定Java开发经验的学习者参考。

图书在版编目(CIP)数据

Android移动应用开发教程/祝永志主编．—北京：清华大学出版社，2018 (2021.12重印)
(21世纪高等学校规划教材·计算机应用)
ISBN 978-7-302-49105-7

Ⅰ．①A… Ⅱ．①祝… Ⅲ．①移动终端－应用程序－程序设计－高等学校－教材
Ⅳ．①TN929.53

中国版本图书馆CIP数据核字(2017)第304166号

责任编辑：郑寅堃
封面设计：傅瑞学
责任校对：胡伟民
责任印制：刘海龙

出版发行：清华大学出版社
网　址：http://www.tup.com.cn，http://www.wqbook.com
地　址：北京清华大学学研大厦A座　**邮　编**：100084
社 总 机：010-62770175　**邮　购**：010-83470235
投稿与读者服务：010-62776969，c-service@tup.tsinghua.edu.cn
质量反馈：010-62772015，zhiliang@tup.tsinghua.edu.cn
课件下载：http://www.tup.com.cn，010-83470236
印 装 者：北京鑫海金澳胶印有限公司
经　销：全国新华书店
开　本：185mm×260mm　**印　张**：16.75　**字　数**：415千字
版　次：2018年5月第1版　**印　次**：2021年12月第6次印刷
印　数：8501～10000
定　价：49.00元

产品编号：069765-01

出版说明

随着我国改革开放的进一步深化，高等教育也得到了快速发展，各地高校紧密结合地方经济建设发展需要，科学运用市场调节机制，加大了使用信息科学等现代科学技术提升、改造传统学科专业的投入力度，通过教育改革合理调整和配置了教育资源，优化了传统学科专业，积极为地方经济建设输送人才，为我国经济社会的快速、健康和可持续发展以及高等教育自身的改革发展做出了巨大贡献。但是，高等教育质量还需要进一步提高以适应经济社会发展的需要，不少高校的专业设置和结构不尽合理，教师队伍整体素质亟待提高，人才培养模式、教学内容和方法需要进一步转变，学生的实践能力和创新精神亟待加强。

教育部一直十分重视高等教育质量工作。2007 年 1 月，教育部下发了《关于实施高等学校本科教学质量与教学改革工程的意见》，计划实施"高等学校本科教学质量与教学改革工程(简称'质量工程')"，通过专业结构调整、课程教材建设、实践教学改革、教学团队建设等多项内容，进一步深化高等学校教学改革，提高人才培养的能力和水平，更好地满足经济社会发展对高素质人才的需要。在贯彻和落实教育部"质量工程"的过程中，各地高校发挥师资力量强、办学经验丰富、教学资源充裕等优势，对其特色专业及特色课程(群)加以规划、整理和总结，更新教学内容、改革课程体系，建设了一大批内容新、体系新、方法新、手段新的特色课程。在此基础上，经教育部相关教学指导委员会专家的指导和建议，清华大学出版社在多个领域精选各高校的特色课程，分别规划出版系列教材，以配合"质量工程"的实施，满足各高校教学质量和教学改革的需要。

为了深入贯彻落实教育部《关于加强高等学校本科教学工作，提高教学质量的若干意见》精神，紧密配合教育部已经启动的"高等学校教学质量与教学改革工程精品课程建设工作"，在有关专家、教授的倡议和有关部门的大力支持下，我们组织并成立了"清华大学出版社教材编审委员会"(以下简称"编委会")，旨在配合教育部制定精品课程教材的出版规划，讨论并实施精品课程教材的编写与出版工作。"编委会"成员皆来自全国各类高等学校教学与科研第一线的骨干教师，其中许多教师为各校相关院、系主管教学的院长或系主任。

按照教育部的要求，"编委会"一致认为，精品课程的建设工作从开始就要坚持高标准、严要求，处于一个比较高的起点上；精品课程教材应该能够反映各高校教学改革与课程建设的需要，要有特色风格、有创新性(新体系、新内容、新手段、新思路，教材的内容体系有较高的科学创新、技术创新和理念创新的含量)、先进性(对原有的学科体系有实质性的改革和发展，顺应并符合 21 世纪教学发展的规律，代表并引领课程发展的趋势和方向)、示范性(教材所体现的课程体系具有较广泛的辐射性和示范性)和一定的前瞻性。教材由个人申报或各校推荐(通过所在高校的"编委会"成员推荐)，经"编委会"认真评审，最后由清华大学出版社审定出版。

目前,针对计算机类和电子信息类相关专业成立了两个"编委会",即"清华大学出版社计算机教材编审委员会"和"清华大学出版社电子信息教材编审委员会"。推出的特色精品教材包括:

(1) 21世纪高等学校规划教材·计算机应用——高等学校各类专业,特别是非计算机专业的计算机应用类教材。

(2) 21世纪高等学校规划教材·计算机科学与技术——高等学校计算机相关专业的教材。

(3) 21世纪高等学校规划教材·电子信息——高等学校电子信息相关专业的教材。

(4) 21世纪高等学校规划教材·软件工程——高等学校软件工程相关专业的教材。

(5) 21世纪高等学校规划教材·信息管理与信息系统。

(6) 21世纪高等学校规划教材·财经管理与应用。

(7) 21世纪高等学校规划教材·电子商务。

(8) 21世纪高等学校规划教材·物联网。

清华大学出版社经过三十多年的努力,在教材尤其是计算机和电子信息类专业教材出版方面树立了权威品牌,为我国的高等教育事业做出了重要贡献。清华版教材形成了技术准确、内容严谨的独特风格,这种风格将延续并反映在特色精品教材的建设中。

清华大学出版社教材编审委员会
联系人:魏江江
E-mail:weijj@tup.tsinghua.edu.cn

前言

面对当前庞大的移动应用开发市场，国内外的 IT 厂商纷纷推出各种移动应用开发平台。Android 是 Google 公司开发的基于 Linux 的开源移动设备操作系统，主要应用于智能手机和平板电脑等移动设备，目前由 Google 倡导成立的开放手机联盟 OHA（Open Handset Alliance）领导开发。Android 已发布最新版本为 Android 7.0。经过几年的快速发展，Android 操作系统在全球得到了大规模的推广，除了应用于智能手机和平板电脑之外，它还可应用于电视、数码相机、游戏机等，可以说目前生活中大多数智能设备都是搭乘 Android 系统设计的。2016 年 11 月，市场研究公司 Gartner 公布的调查报告显示，在过去的一个季度中，苹果售出 4300 万部 iPhone，而 Android 销售量则达到了 3.28 亿部，Android 占到过去一个季度所售出智能手机的 88%，而 iOS 市场份额仅仅高于 10%。由于 Android 迅速发展，使得市场对 Android 开发人才的需求激增，因此学好 Android 开发技术将会使读者在更广阔的人才市场竞争中赢得先机。目前，关于 Android 开发应用的书籍已经很多，但是适合作为高等院校教材的却很少。为了满足对 Android 应用开发教材的需求，我们在多年理论教学、应用开发的基础上，不断总结教学经验，围绕 Android 开发新技术，编写了本书。

读者对象

本书适合于从事 Android 应用开发的初、中级人员。根据多年的教学体会和实际开发经验，我们慎重地安排了本书的内容。从移动信息设备平台、Android 的架构及 Android 开发环境搭建入手，到有一定深度的 UI 控件及布局设计技术；从 Activity、Intent、Service 到数据存储与网络通信技术的阐述，本书为读者从事 Android 应用开发提供了基础而又全面的内容，提供了大量从实际开发中提炼出来的应用案例，有的案例读者甚至不加修改就可以用于自己的开发项目中。通过学习本书，读者不但能掌握 Android 应用开发的基本步骤，还能培养学以致用的专业素养。

本书结构

全书共 9 章。

第 1 章 Android 系统概述，讲述 Android 移动应用开发的基本知识，主要介绍移动信息设备的平台、Android 的基本概念以及 Android 应用的基本构成等。

第 2 章开发环境的搭建，讲述 Android 应用开发环境的搭建，主要讲述 Android 开发环境的安装、配置，包括安装 JDK 及配置环境变量；安装 Eclipse、安装 Android Studio 以及 Android 程序的一些调试工具；介绍了 Eclipse 环境与 Android Studio（AS）环境之间的转换与不同环境之间的比较。

第 3 章 Activity 及其生命周期，讲述 Activity 的创建、Activity 的生命周期及其案例、JUnit 测试以及资源调用等。

第 4 章常见的 UI 控件，讲述 Android 基本控件的使用方法，包括 TextView、EditText、Button、ImageView 等；常见的弹出框基本使用，包括 ProgressBar、AlertDialog、ProgressDialog 以及 Toast 等；Listview 的基本使用、自定义控件、引用布局以及创建自定义布局等。

第 5 章 Intent 与组件通信，讲述 Intent 启动组件的方式；隐式 Intent 及 Intent 相关属性，包括 Component（组件）、Action（动作）、Category（类别）、Data（数据）、Type（数据类型）、Extras（扩展信息）、Flags（标志位）等；隐式 Intent 的具体应用，包括打开指定网页、打电话、发送短信、播放指定路径音乐、卸载程序、安装程序，以及向下一个应用传递数据、返回等。

第 6 章 Android 后台服务，讲述 Service 的基本用法，包括创建、配置 Service，启动 Service、Service 和 Activity 进行通信等；Service 的生命周期等；Service 其他用法，包括使用前台服务、使用 IntentService 等；常见的系统服务，包括电话管理器、短信管理器、振动器、闹钟/全局定时器等。

第 7 章数据存储，主要讲述 Android 操作系统为数据存储提供的五种方式：使用文件存储（File 存储）、首选项存储（Preferences 存储）、数据库存储（SQLite 存储）、内容提供者（Content Providers）以及网络存储（NetWork）等。

第 8 章网络通信，讲述 Socket 通信，包括 Socket 客户端的开发、Socket 服务器端的开发以及简单聊天室等；基于 HTTP 的网络编程，包括 HttpURLConnection 的使用方法、HttpClient 的使用方法等；基于 WebView 的网络编程，包括 WebView 视图组件以及使用 WebView 浏览网页等。

第 9 章移动办公软件系统，为综合案例部分，讲述项目架构，日期和时间、定位、天气三大功能等，通知公告模块、工作日志模块、考勤管理模块、费用申请模块、请假模块和设置模块六大模块等。

本书由祝永志主编，第 1～3 章由祝永志和朱盼盼共同完成，第 4～7 章由祝永志和申健共同完成，第 8 章由祝永志和刘梦芸共同完成，第 9 章综合案例部分由祝永志、申健调试开发。本书的所有例子程序全都经过测试，读者可放心使用。全书 Android 程序开发环境是 Eclipse，也可以在开发环境 Android Studio 中调试运行。

由于作者水平有限，对书中不足之处，欢迎广大读者和同行指正。

联系作者

欢迎读者通过电子邮箱 rizhaozyz@126.com 与作者取得联系。本书的 PPT 课件和书中程序源代码可到清华大学出版社网站上下载。

致谢

感谢《Android 程序设计》（青岛东合信息技术有限公司编著）及《Android 移动应用基础教程》（传智播客高教产品研发部编著）等优秀书刊，感谢网络平台上许多 Android 资料，本书作者从中获取了不少有价值的信息，在此向这些资源作者表示衷心的感谢，并以此书向他们表达我们的敬意。

编　者

2017 年 8 月

目 录

第1章 Android系统概述

本章重点

- 移动信息设备的平台
- Android 的基本概念
- Android 应用的基本构成

网络时代，智能设备走进了人们的生活。手机已不仅仅是打电话的工具，微信、支付宝、QQ 等各种 APP 极大地丰富和方便了人们的生活。智能手环可以检测人的运动、睡眠等情况，并可通过一些 APP 给人们提出健康生活的建议。

1.1 移动信息设备的平台

中国移动设备的演变历程，是中国科技发展的一个缩影。

从 20 世纪 80 年代开始，中国步入了移动通信时代，移动通信设备如雨后春笋般涌现，例如能够“移动着接听”的大哥大，“别在腰上”的 BP 机，“手机、呼机、商务通，一个都不能少”所宣传的商务通 PDA，红极一时的小灵通，等等。

人们记忆中的移动设备，如图 1-1 所示。

图 1-1 人们记忆中的移动设备

当年这些移动设备都曾经风光地存在过，但是随着科技的发展都默默地离开了历史的舞台。如今，我们已经走进了智能设备时代。

可以搭载智能平台的设备多种多样，如手机、平板电脑、智能手环、智能电视等。

1.1.1 移动通信设备的操作系统

移动通信设备中的操作系统市场呈现出群雄割据的局面，那些用过或者出现在记忆中的操作系统有：

- Symbian
- Windows Mobile
- iOS
- Android
- BlackBerry OS

由于采用的技术或者开发语言不同，这些系统之间的应用软件互不兼容，因此不同的应用需要考虑不同的软、硬件搭配和标准不一的操作系统。

目前，各大移动设备厂商对市场虎视眈眈，从 2010 年开始移动互联网进入了一个新的快速发展时期。

1. 智能手机操作系统之 Symbian（塞班）

Symbian 是一家研发与授权 Symbian 操作系统的软件公司。全球各大手机领导厂商，例如摩托罗拉、诺基亚、三星、西门子与索尼爱立信等都曾经使用过 Symbian 操作系统。虽然 Symbian 作为昔日智能手机操作系统的王者已经淡出了历史舞台，但作为智能手机操作系统的先驱者之一，还是应简单地介绍一下。

在 2005～2010 年期间，智能机市场上占有率曾一度领先，可以看到很多人都在使用 Symbian 系统的手机。但作为在智能手机市场上曾经占据领先地位的手机操作系统，它的繁荣可以说是昙花一现。

诺基亚公司在 2011 年 12 月 21 日宣布放弃 Symbian 品牌，这标志着 Symbian 系统已走向没落。

诺基亚于 2014 年 1 月 1 日正式停止了 Nokia Store 应用商店内对 Symbian 应用的更新，也禁止开发人员发布新应用，这标志着彻底告别 Symbian 系统。

图 1-2　苹果手机

2. 智能手机操作系统之 iOS

iOS 是苹果公司开发的一款手机操作系统，它主要应用于其旗下的 iPhone 系列手机，同时也可以用于苹果公司的其他系列产品，如 iPod、iTouch 以及 iPad 等。苹果手机如图 1-2 所示。

iOS 的系统架构分为四个层次：核心操作系统层（the Core OS Layer）、核心服务层（the Core Services Layer）、媒体层（the Media Layer）和可轻触层（the Cocoa Touch Layer）。

2015 年 6 月 8 日苹果发布了全新的 iOS 9 操作系统，它具有分屏操作、画中画等让人眼花缭乱的功能。2016 年 12 月 12 日苹果发布最新 iOS 10.2。

3. 智能手机操作系统之 Android

2008 年 Google 公司发布了开源手机操作系统——Android。它的诞生，标志着移动信息设备操作系统的发展进入一个崭新的阶段。

该平台由四个层次组成，它们分别是 Linux 内核层（Linux Kernel）、系统运行时库层（Libraries 和 Android Runtime）、应用程序架构层（Application Framework）以及应用程序层(Applications)。

Android 的作用不仅是手机操作系统,还是开源的,是由开放手机联盟共同支持的首个移动软件开发平台。

Android上APP的开发语言是Java,并且谷歌公司专门为其提供了开发所使用的SDK。2016年上半年,Android操作系统全球市场占有率为86.2%,处于绝对领先地位。Android手机如图1-3所示。

图1-3 Android手机

4. 智能手机操作系统之 Windows Phone

2008年,微软公司在目睹iOS和Android在智能手机领域攻城略地后,重新建立了Windows Mobile小组来开发微软公司自己的手机操作系统。

微软公司想要通过全新的Windows手机,把网络、个人计算机和手机的优势集于一身,通过让人们可以随时随地使用Windows来方便人们的生活。Windows Phone内置了Office办公套件和Outlook等在传统个人计算机领域使用的办公软件,使得人们通过手机依然可以更加有效和方便地办公。

图1-4 Windows Mobile 手机

在应用方面,Windows Phone提供了很好的开发工具。

Windows Phone的应用数量很少,其界面使用了磁贴的设计,使得图标看上去千篇一律,容易造成审美疲劳。虽然在Windows Phone 7.5之后的版本开始支持多任务处理,但是最多也只能运行5个程序,在这一点上远输于Android和iOS。Windows Mobile手机如图1-4所示。

2015年,它的市场份额持续下降至1.7%,2016年微软公司推出Windows 10 Mobile系统。

5. 智能手机操作系统之 BlackBerry OS(黑莓)

BlackBerry OS是加拿大RIM(Research In Motion)为其智能手机产品BlackBerry开发的专用操作系统。

BlackBerry OS支持移动式电子邮件、移动电话、网页浏览、文字短信、互联网传真及其他通信和互联网服务。

BlackBerry的开发平台分为三部分,分别是BlackBerry Browser Development(黑莓浏览器开发)、Rapid Application Development(快速程序开发)和Java Application Development(Java程序开发)。

BlackBerry同时支持标准Java ME程序和开发黑莓专用的Java程序。

2014年11月14日,黑莓官方宣布,与三星及其他知名科技公司达成协议,将要进一步拓展其应用领域。BlackBerry手机如图1-5所示。

图1-5 BlackBerry手机

数据调查机构 NetApplications 报告显示，2016 年上半年全球范围内黑莓市场份额已经降至 0.85%，详见图 1-6。

OPERATING SYSTEM	TOTAL MARKET SHARE
Android	66.01%
iOS	27.84%
Windows Phone	2.79%
Java ME	1.44%
Symbian	1.03%
BlackBerry	0.85%
Samsung	0.01%
Kindle	0.01%
Bada	0.00%
LG	0.00%
Windows Mobile	0.00%

图 1-6　移动操作系统市场份额统计

1.1.2　开放手机联盟

2007 年 11 月 5 日，美国 Google 公司宣布组建的一个全球性的联盟组织——开放手机联盟(Open Handset Alliance，OHA)支持 Android 的手机操作系统。开放手机联盟最初包括手机芯片厂商、手机制造商、移动运营商在内的 34 家成员。

移动手机联盟创始成员有高通、三星、SiRF、SkyPop、Sonic Network、Sprint Nextel、Aplix、Ascender、Audience、Broadcom、中国移动、eBay、Esmertec、谷歌、宏达电、英特尔、KDDI、Living Image、LG、Marvell、摩托罗拉、NMS、NTT DoCoMo、Nuance、Nvidia、PacketVideo、Synaptics、TAT 、意大利电信、西班牙电信、得州仪器、T-Mobile 和 Wind River。

截至 2012 年，开放手机联盟的成员已达到 84 家。

1.1.3　4G 时代来临

移动互联网，是将移动通信和互联网二者结合为一体，使人们可以更加方便快捷地享受移动通信技术和互联网技术所带来的福利。

第四代移动通信技术(The Fourth Generation，4G)时代已经开启。

移动终端设备已经为移动互联网的发展注入巨大的能量，依靠 4G 技术的发展，P2P、O2O 等新名词已经和人们的生活密不可分，移动互联网产业一定会迎来前所未有的发展。

2016 年 1 月 7 日，国内移动数据服务商 QuestMobile 发布了《2015 年中国移动互联网研究报告》。报告称：截至 2015 年 12 月，国内在网活跃的移动智能设备数量已经达到 8.99 亿。报告分析称，苹果设备与 Android 设备持有量比例为 3∶7，其中使用移动通信设备用户的男女比例大约是 6∶4。

手机以其便捷性已经取代个人计算机成为第一大上网终端。我国移动互联网发展已经进入全民移动互联网时代。

从 2016 年开始，移动互联网已迎来崭新的时代，移动电子商务、视频营销、移动手机支付、O2O、大数据、4G 手机网站、QQ 营销、微信营销、微商城、微官网、云营销、娱乐圈营销、物联网、实用 APP 等各种服务在潜移默化地改变着人们的生活方式。

随着科学技术的发展，移动客户端开发将更加便捷化。基于大数据、云平台的运营管理，数字化、智能化的生活方式的转变，都将为移动互联网开发平台提供有力的支撑和原动力。

1.2　Android 的介绍

Android 一词原本是法语中的“机器人”，谷歌公司使用一个绿色的小机器人作为 Android 的标识。

1.2.1　Android 的发展史

Android 系统是最初由安迪·鲁宾（Andy Rubin）于 21 世纪初创立的手机操作系统，2005 年被谷歌收购。

Android 是一个以 Linux 为基础的开源移动设备操作系统，主要安装在智能手机和平板电脑上，由 Google 公司成立的 OHA 领导开发。截至目前，Android 发布的最新版本为 Android 7.0。

从 Android 1.5 开始，Android 使用甜点作为系统版本代号，如表 1-1 和图 1-7 所示。

表 1-1　Android 操作系统各版本发布时间表

Android 版本	发布日期	代　号
Android 1.1	2008 年 9 月	
Android 1.5	2009 年 4 月 30 日	Cupcake（纸杯蛋糕）
Android 1.6	2009 年 9 月 15 日	Donut（炸面圈）
Android 2.0/2.1	2009 年 10 月 26 日	Eclair（长松饼）
Android 2.2	2010 年 5 月 20 日	Froyo（冻酸奶）
Android 2.3	2010 年 12 月 6 日	Gingerbread（姜饼）
Android3.0/3.1/3.2	2011 年 2 月 22 日	Honeycomb（蜂巢）
Android 4.0	2011 年 10 月 19 日	Ice Cream Sandwich（冰淇淋三明治）
Android 4.1	2012 年 6 月 28 日	Jelly Bean（果冻豆）
Android 4.2	2012 年 10 月 8 日	Jelly Bean（果冻豆）
Android 5.0	2014 年 10 月 15 日	Lime Pie（酸橙派）
Android 6.0	2015 年 5 月 28 日	Marshmallow（棉花糖）
Android 7.0	2016 年 5 月 18 日	Nougat（牛轧糖）

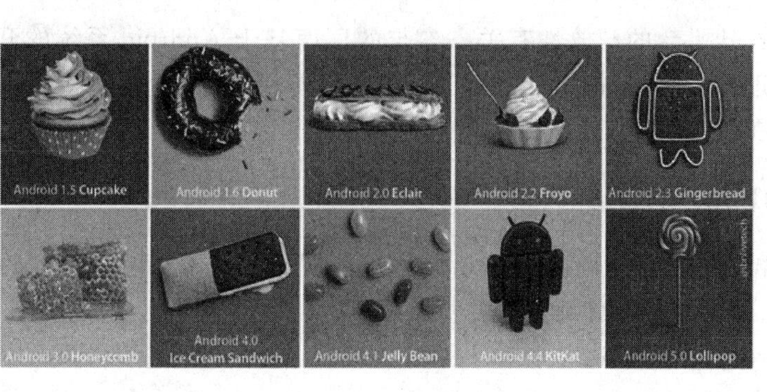

图 1-7　Android 各版本的标识

Android 6.0 的主要新功能有如下几点。

(1) 锁屏语音搜索。Android 6.0 加入了锁屏状态下的语音搜索功能,使得用户可以在锁屏状态下直接进行语音搜索而不需要解锁屏幕再进行语音搜索。

(2) 指纹识别。Android 6.0 加入指纹识别的功能。Android 6.0 提供了原生的指纹识别 API,这样减少了手机厂商独自开发指纹识别模块的成本。原生指纹识别功能会提升 Android 智能手机通过指纹识别进行支付的安全性。

(3) 电量管理。Android 6.0 拥有自带 Doze 电量管理功能。在这一模式下,手机会检测应用的使用情况,检测到应用一段时间未使用时,就会清理后台进程进而减少功耗。

(4) App Links。Android 6.0 通过 App Links 功能能够向网络服务器提出申请,进而自主识别链接内容。

(5) Android Pay。Android 6.0 还加入了 Android Pay 功能。这一功能强化移动支付体验。这也是为了对抗 Apple Pay 加入的新功能。

Android 7.0 的主要新功能有如下几点。

(1) 分屏多任务。Android 7.0 原生支持分屏多任务,用户单击多任务按键后,长按一个应用程序图标,将它拖到屏幕顶部或者底部,再去单击另一个应用程序图标,即实现了分屏多任务。

(2) Data Saver。当 Data Saver 功能开启后,黑名单中的 APP 将会受到流量限制。

(3) 画中画模式。Android 7.0 画中画功能与 iOS 9 的画中画基本一致,只不过 Android 主要是针对电视平台的。

1.2.2 Android 优缺点

对手机厂商来说,Android 的优势有以下几个方面。

1. 开放性

Android 系统最大的优势就是其开放性。作为一个开源平台,Android 允许任何硬件公司、应用开发团队、通信企业加入到 Android 联盟中来。基于以上特点,越来越多 Android APP 开发者加入到这一行列。这一原则对于 Android 的初期发展而言,有利于吸引更多资源投入到这一阵营中,为 Android 带来更多的人气。

2. 丰富的硬件选择

硬件丰富性与 Android 平台的开放性有关。参与的厂商基于 Android 系统的开放性,会根据不同的需求以及理念的不同,推出具有自己特色的设备和系统,例如 MIUI 等。这些功能上的差异和特色处在较高的层次上,不会影响软件的兼容性。

3. 不受任何限制的应用开发商

Android 平台提供给第三方开发商一个十分宽泛、自由的环境,不会受到各种条条框框的限制,因此会有无数新颖别致的软件诞生。

对开发者来说 Android 的优势有:

(1) 源代码免费开放;

(2) 开发工具廉价；

(3) 发布模式方便；

(4) 盈利方式多样。

Android 的不足之处有：

(1) 安全和隐私。许多的手机应用会偷偷地读取位置信息、通话记录、短消息等隐私信息，在手机主人不知情的情况下发送给服务器，这大大地影响了人们手机使用的安全性。

(2) 运营商预装手机应用。在国内市场，三大电信运营商的定制机会预装不少的 APP，尤其是运营商本身的业务，引得不少顾客对定制机不满，这大大影响了顾客的体验和手机的流畅性。

(3) 产品代数太多。由于 Android 版本繁多，一些早期的版本并没有退出使用，这使得应用开发者需要兼顾多个 Android 的版本，增加了开发成本和时间。

1.2.3 Android 盈利方式

1. 应用内收费类

这种模式比较适合游戏或者服务类的应用。一般而言，有两类收费模式。第一类是 Android 应用本身免费，靠虚拟货币或者道具盈利，例如一些网络游戏；第二类是基本功能免费，升级或者高级功能(例如高级会员)收费，例如 QQ 等。

Android 应用中，可以借助合理地设定收费模式以及跟第三方支付平台的对接来实现。这种模式具有较高的收益转化率，但这要通过开发出能吸引人的应用或服务来实现。

2. 广告收入

应用免费，靠广告盈利，即在游戏或 Android 应用运行中，向玩家展示广告。有展示、单击和注册三种方式来得到收益。

3. 委托开发

一些企业或者商家会向开发者支付一定费用来定制自己的 Android 应用。开发者可以获得一定报酬，并对其提供后续的服务或者应用版本的演进。

1.3 Android 的架构

Android 系统采用层次化系统架构，官方公布的标准架构如图 1-8 所示。这种软件叠层的架构由低到高分为四个主要功能层，分别是 Linux 内核层(Linux Kernel)、系统运行时库层(Libraries 和 Android Runtime)、应用程序框架层(Application Framework)以及应用程序层(Applications)。

1. Linux 内核层

Android 系统主要基于 Linux 内核开发。Linux 内核层为 Android 的各种硬件设备提供底层的驱动，包括显示驱动、USB 驱动、照相机驱动、键盘驱动、蓝牙驱动、WiFi 驱动、

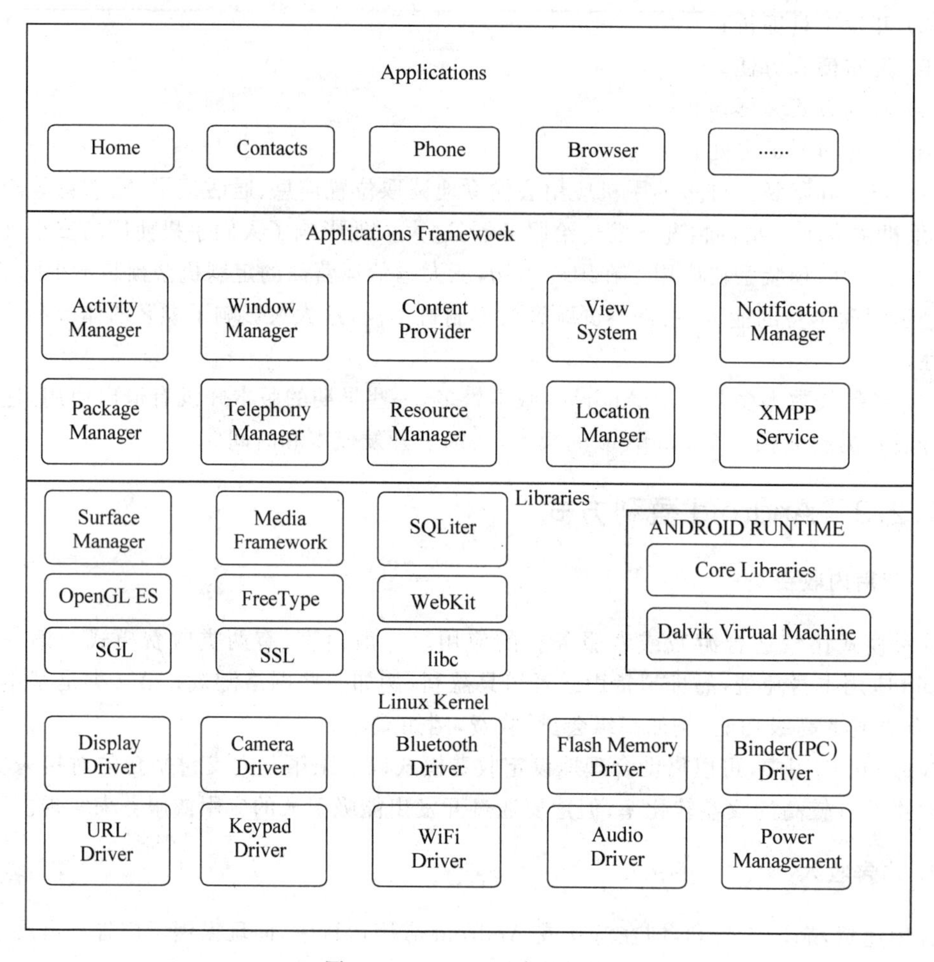

图 1-8 Android 平台的架构

M-System 驱动、声卡驱动、Binder 驱动以及电源管理驱动等。服务实现硬件设备驱动、进程和内存管理、网络协议栈、电源管理、无线通信等核心功能。

2. 系统运行时库层

在图 1-8 中，位于 Linux 内核层之上的系统运行时库层是应用程序框架的支撑，为 Android 系统中的各个组件提供服务。系统运行时库层由系统类库和 Android 运行时库构成。

(1) 系统类库。系统类库大部分由 C/C++编写，通过 C/C++库为 Android 系统提供主要的特性支持。这一层的功能包括多媒体库、WebKit、SGL(Skia Graphics Library)、媒体框架、OpenGL ES(OpenGL for Embedded Systems)和 SQLite 等。

(2) Android 运行时库。这一层包括 Java 核心库和 Android 虚拟机两部分。Java 核心库包含提供 Java 编程语言核心库的大部分功能和 Android 的核心库。Android 的 Dalvik 虚拟机类似于 Java 虚拟机，它是专门为移动设备设计的，其特点是用最少的内存资源来运行代码，并且能同时执行多个虚拟机。它所运行的文件不能通过 Java 代码编写，需要通过 Android SDK 进行转换。

3. 应用程序框架层

Android 应用程序框架如表 1-2 所示，主要提供构件应用程序可能需要的各种 API。该应用程序框架简化成组件的重用，使开发人员可以进行快速的应用程序开发，可以通过继承实现个性化的扩展。基于应用程序框架层发布的功能模块可以被其他应用程序所调用。

表 1-2　Android 应用程序框架

应用程序框架层	功　　能
活动管理器(Activity Manager)	管理各个应用程序的生命周期，并且提供常用的导航回退功能
窗口管理器(Window Manager)	对所有开启的窗口程序进行管理
内容提供器(Content Provider)	实现应用程序之间数据共享的有效途径
视图系统(View System)	包括列表(Lists)、网格(Grids)、文本框(Text Views)、按钮(Buttons)、可嵌入的 Web 浏览器
通知管理器(Notification Manager)	使得应用程序可以在状态栏中显示自定义的客户提示信息
包管理器(Package Manager)	对应用程序进行管理，提供安装应用程序、卸载应用程序以及查询相关权限信息等功能
资源管理器(Resource Manager)	提供非代码资源的使用，例如本地化字符串、图片、音频和布局文件等
位置管理器(Location Manager)	提供位置服务
电话管理器(Telephony Manager)	管理所有的移动设备功能
XMPP 服务(XMPP Service)	是 Google 在线即时交流软件中一个通用的进程，提供后台推送服务

4. 应用程序层

Android 平台的应用程序层是各种应用软件，包括智能手机上实现的常见基本功能程序，如短信、电话、图片浏览器、日历、游戏、地图、Web 浏览器等程序。这些应用程序都是使用 Java 语言编写的。

本书讲解的 Android 开发的内容主要是基于 Android 平台的应用层。

1.4　本章小结

本章作为全书的开篇，主要介绍了移动通信设备操作系统、开放手机联盟和 Android 的概念、发展历程现状及其优缺点，介绍了 Android 架构的四层结构体系。

本章讲授的内容是 Android 中基础性知识，要求初学者必须掌握。从第 2 章开始，将学习具体的 Android 应用程序开发技术。

1.5　练习题

一、填空题

1. Android 操作系统是由________于________年开发的。
2. Android 操作系统是基于________系统开发的。

3. 开放手机联盟是由________公司主导，于________年成立的。

4. iOS 是________公司开发的手机操作系统，分为________、________、________、________四层。

二、选择题

1. 以下不属于Android操作系统架构层次的是(　　)。

A. Linux 内核层　　B. 系统运行时库层

C. 应用程序架构层　　D. 可轻触层

2. 下列不属于应用层的是(　　)。

A. 相机　　B. 短信　　C. 电话　　D. 音频驱动

3. 下列企业不属于开放手机联盟的是(　　)。

A. 微软　　B. 谷歌　　C. 三星　　D. 小米

三、简答题

1. Android 平台的技术架构分为哪几部分？

2. 简述 Android 市场份额为何会超过 iOS。

四、编程题

编写一个 Android 应用程序，在屏幕上显示“好好学习 Android，努力成为软件开发大师!”。

第2章 开发环境的搭建

本章重点

- 开发环境的安装与配置
- 熟悉开发环境
- 不同环境之间的转换

在今后学习和开发 Android 应用的过程中，首先需要在自己的计算机上编写代码和进行测试，然后才是将其部署到真实设备上进行各种测试。

在本章中，你将学习如何搭建和配置 Android 的开发环境。在这个过程中，可以熟悉开发环境，学习简单操作，这是将来应用开发的基础。由于 Android 开发环境多种多样，本章将介绍 Eclipse 环境以及 Android Studio 环境的搭建，并介绍如何将 Eclipse 的项目迁移到 Android Studio 中。

2.1 开发环境的安装与配置

Android 应用程序是用 Java 语言开发的，因此可以使用 Java 开发工具 Eclipse 通过安装合适的 ADT(Android Developer Tools)和 SDK(Software Development Kit)进行开发。

Android Studio 是 2013 年 5 月 Google 在 Google I/O 大会上发布的全新开发 Android 的 IDE。它是基于 IntelliJ IDEA 的。在 2014 年 12 月 Google 发布了第一个稳定版(1.0)。Google 官方将逐步放弃对原来主要的 Eclipse ADT 的支持，并为 Eclipse 用户提供工程迁移的解决办法。2015 年初 Google 发布了最新版本 Android Studio 2.0 RC 3。Android Studio 依托 IntelliJ IDEA 开发，与 Eclipse 相比它更智能，提示功能更加强大，默认使用 Gradle 构建，拥有布局文件实时预览等功能。

2.1.1 安装 JDK 及配置环境变量

Android 应用程序是用 Java 语言开发的，因此需要在计算机上安装、配置 JDK(Java Development Kit)。

JDK 是专门为 Java 语言设计的软件开发工具包。移动设备上的应用程序开发主要使用 Java 语言，具有跨平台的特性。JDK 包含了 Java 的运行环境、Java 工具和 Java 基础的类库。这些是使用 Java 语言开发必不可少的。

1. 到官网下载 JDK8

可以在 Oracle 官网下载最新版的 JDK。下载网址如下：

http://www.oracle.com/technetwork/Java/Javase/downloads/index.html

在官网选择下载 JDK，如图 2-1 所示。选择接受协议单选按钮来接受许可证协议。然后选择适合自己操作系统的 JDK，如图 2-2 所示。

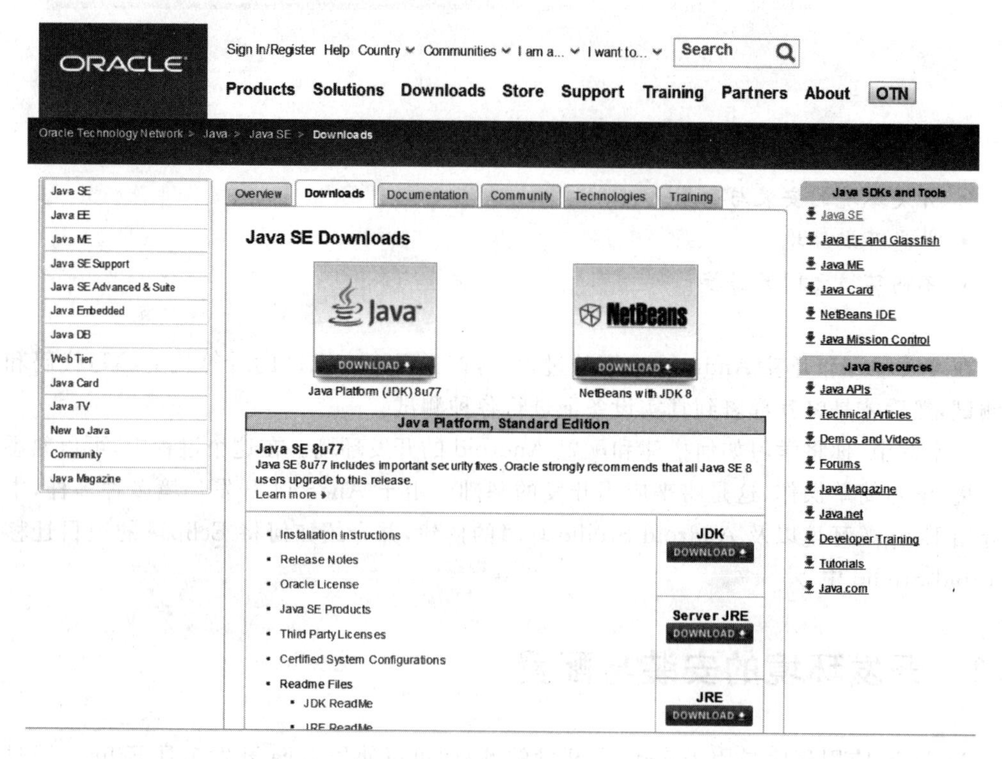

图 2-1 在官网选择下载 JDK

Java SE Development Kit 8u77

You must accept the Oracle Binary Code License Agreement for Java SE to download this software.

○ Accept License Agreement ◉ Decline License Agreement

Product / File Description	File Size	Download
Linux ARM 32 Soft Float ABI	77.7 MB	jdk-8u77-linux-arm32-vfp-hflt.tar.gz
Linux ARM 64 Soft Float ABI	74.68 MB	jdk-8u77-linux-arm64-vfp-hflt.tar.gz
Linux x86	154.74 MB	jdk-8u77-linux-i586.rpm
Linux x86	174.92 MB	jdk-8u77-linux-i586.tar.gz
Linux x64	152.76 MB	jdk-8u77-linux-x64.rpm
Linux x64	172.96 MB	jdk-8u77-linux-x64.tar.gz
Mac OS X	227.27 MB	jdk-8u77-macosx-x64.dmg
Solaris SPARC 64-bit (SVR4 package)	139.77 MB	jdk-8u77-solaris-sparcv9.tar.Z
Solaris SPARC 64-bit	99.06 MB	jdk-8u77-solaris-sparcv9.tar.gz
Solaris x64 (SVR4 package)	140.01 MB	jdk-8u77-solaris-x64.tar.Z
Solaris x64	96.18 MB	jdk-8u77-solaris-x64.tar.gz
Windows x86	182.01 MB	jdk-8u77-windows-i586.exe
Windows x64	187.31 MB	jdk-8u77-windows-x64.exe

图 2-2 选择适合自己操作系统的 JDK

2. 安装 JDK

下载合适版本的 JDK,双击图标进行 JDK 的安装。

计算完空间要求后,单击"下一步"按钮,如图 2-3 所示。

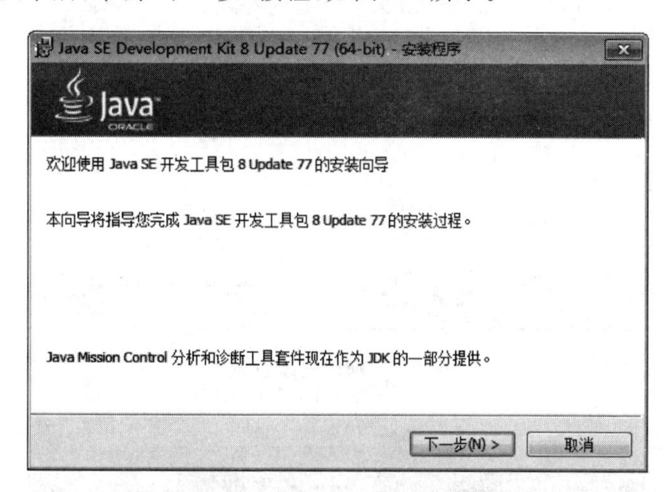

图 2-3 单击"下一步"按钮

如图 2-4 所示,单击"更改"按钮,可以设置要安装 JDK 的位置。

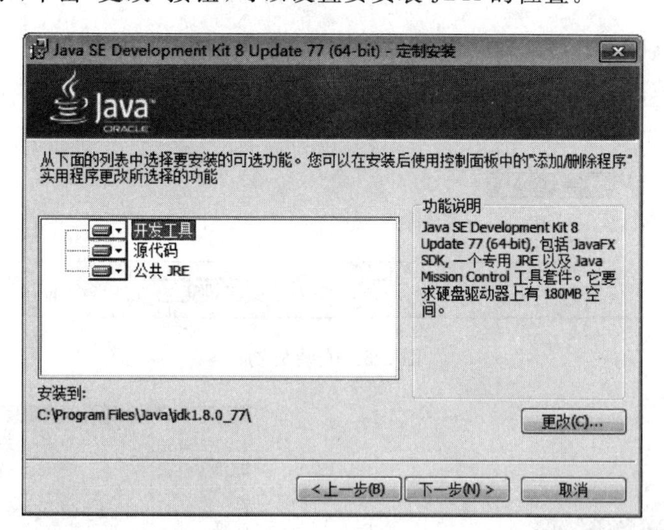

图 2-4 单击"更改"按钮,可设置安装 JDK 的位置

如图 2-5 所示,单击"下一步"按钮进行 JDK、JRE 的安装。

如图 2-6 所示,单击"关闭"按钮,安装完成。

3. 配置环境变量

鼠标右击"我的电脑",在弹出的菜单中选择"属性",然后,在弹出的"系统"窗口中,选择"高级系统设置",之后在"系统属性"对话框的"高级"选项卡中单击"环境变量"按钮,如图 2-7 所示。

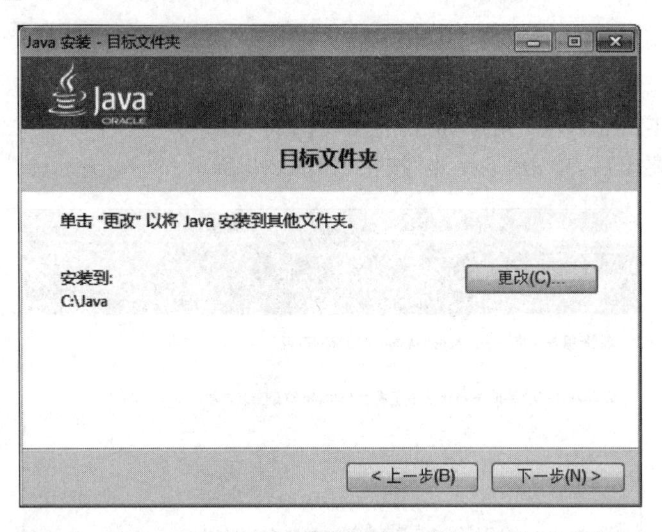

图 2-5 确定安装位置

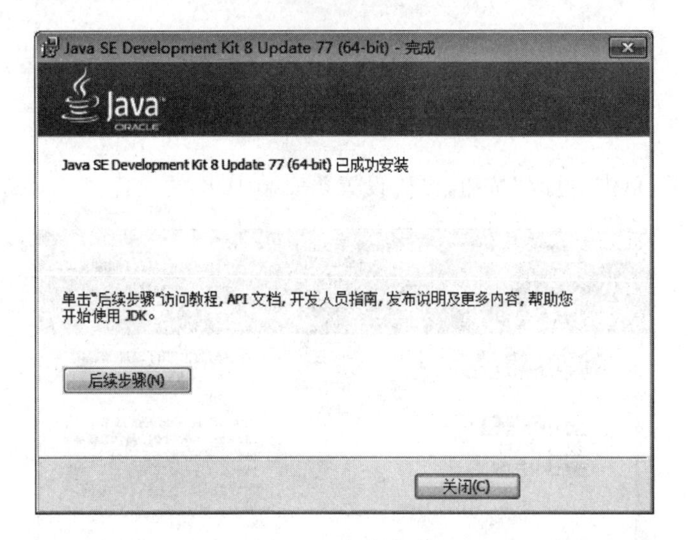

图 2-6 安装成功

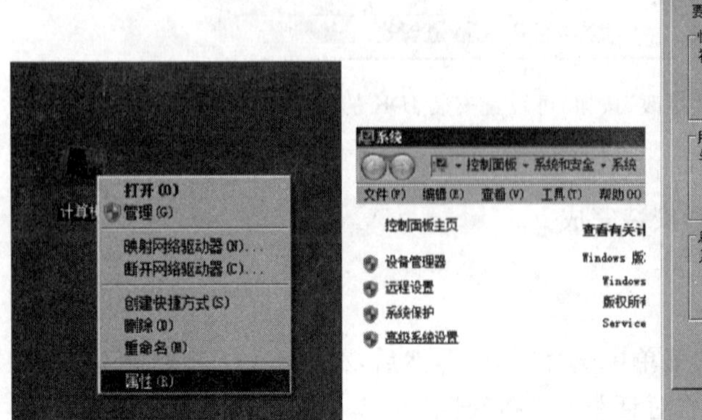

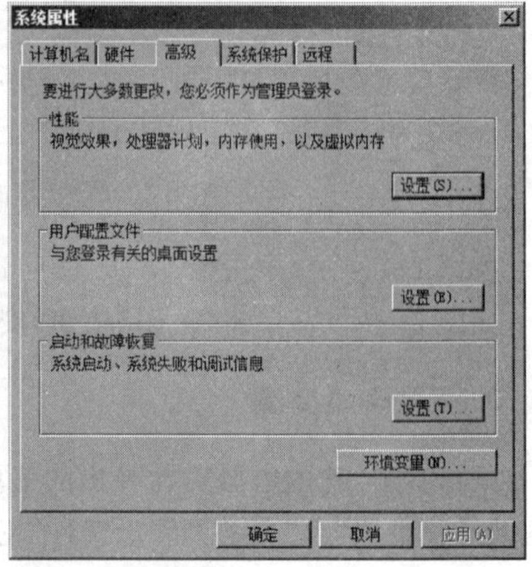

图 2-7 单击“环境变量”按钮

如图 2-8 所示，在“环境变量”对话框中单击“系统变量”下的“新建”按钮，新建变量 JAVA_HOME，把变量值设置为安装 JDK 的路径；在“系统变量”下新建或编辑 Path 变量，把变量值设置为 JDK 的 bin 路径（在 JDK 1.7 以后不必设置 classpath 路径），Path 变量添加上“%JAVA_HOME%\bin;%JAVA_HOME%\jre\bin;”，注意要使用分号进行分隔。

图 2-8 “环境变量”设置

之后需要检查 JDK 是否安装成功。在“开始”菜单的“搜索程序和文件”框中输入 cmd 打开命令行，在命令行中输入“Java -version”来验证是否安装成功。如果出现版本号即表示安装成功。

2.1.2 安装 Eclipse 环境

1. 下载 Eclipse

Android 应用程序是用 Java 语言进行开发的，因此可以在计算机上安装配置 Eclipse 来进行开发。

可以在 Eclipse 官网下载最新版的 Eclipse。下载网址如下：

```
http://www.eclipse.org/downloads/
```

在官网选择下载 Eclipse，如图 2-9 和图 2-10 所示。

2. 解压 Eclipse

完成解压后如图 2-11 所示，将 eclipse.exe 建立桌面快捷方式，双击桌面上的 eclipse 图标，启动的时候需要选择工作空间的路径，配置完成后单击“OK”按钮就可以运行了。

3. 安装 ADT 插件

选择菜单栏中的 Help|Install New Software，在弹出的窗口中单击“Add”按钮，如图 2-12 所示。

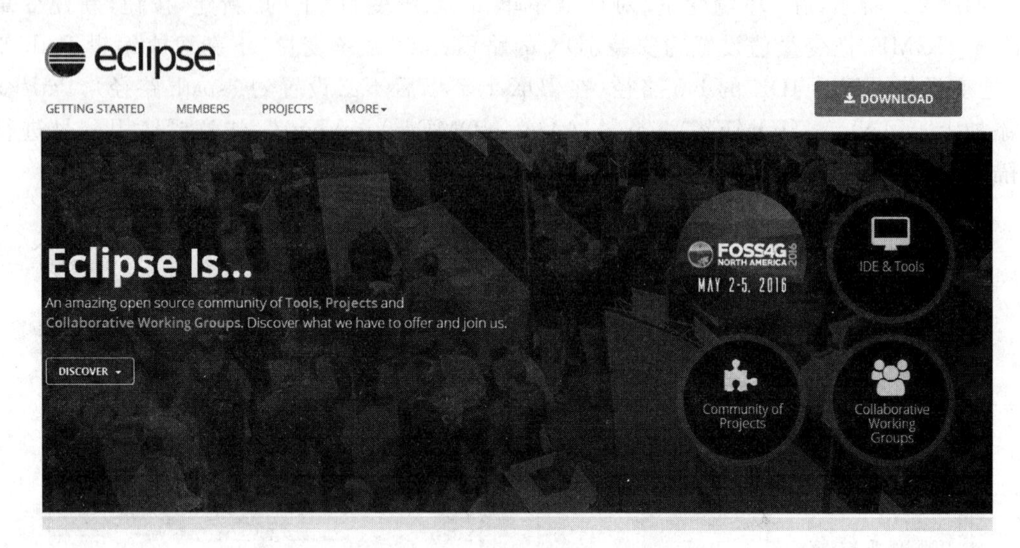

图 2-9　Eclipse 官网

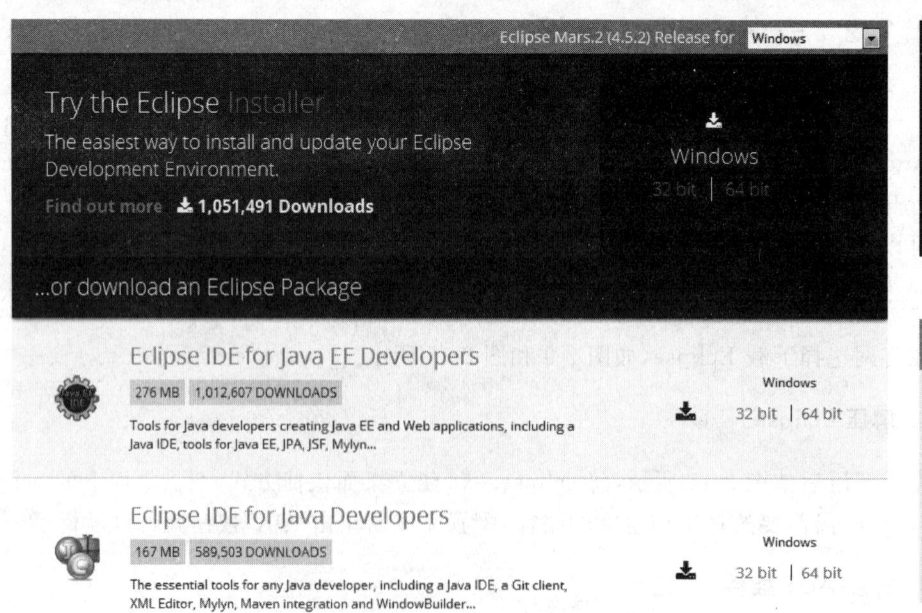

图 2-10　下载最新版的 Eclipse

图 2-11 解压后的文件

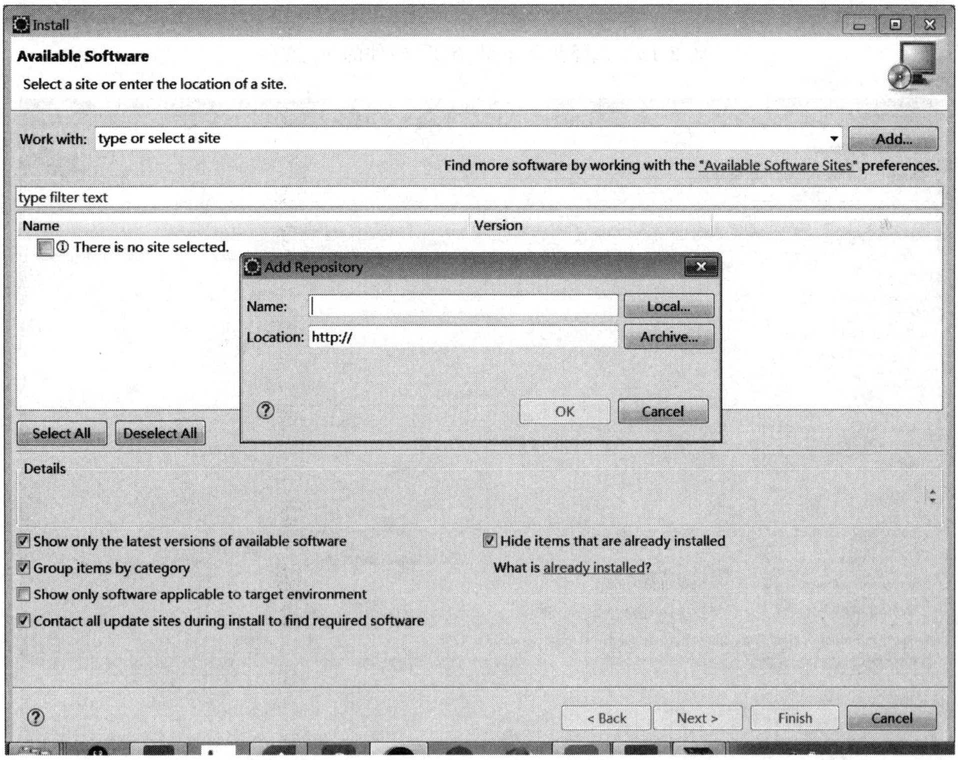

图 2-12 Install 窗口

单击“Archive…”按钮，选择想要安装的 ADT 插件的 zip 文件。选择完成后单击“OK”按钮。勾选想要安装的内容，如图 2-13 所示。单击“Next”按钮，开始 ADT 的安装，如图 2-14 所示。安装完成，如图 2-15 所示，单击“Next”按钮。进入使用许可界面，选择同意。插件安装完成后，Eclipse 会提示重启。重启完成后 ADT 就生效了。

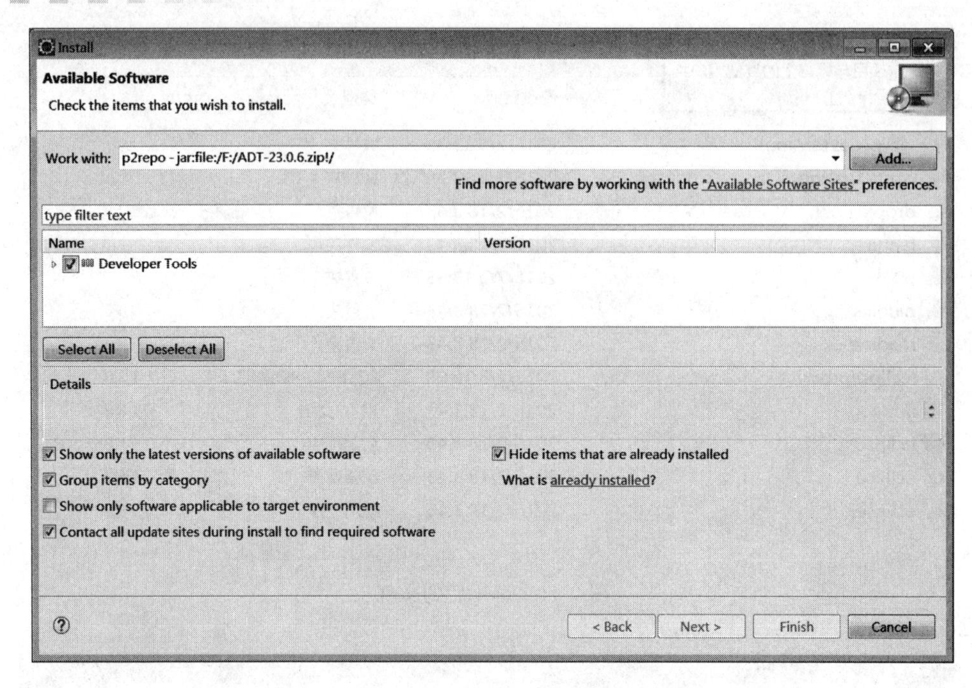

图 2-13　选择想要安装 ADT 插件的 zip 文件

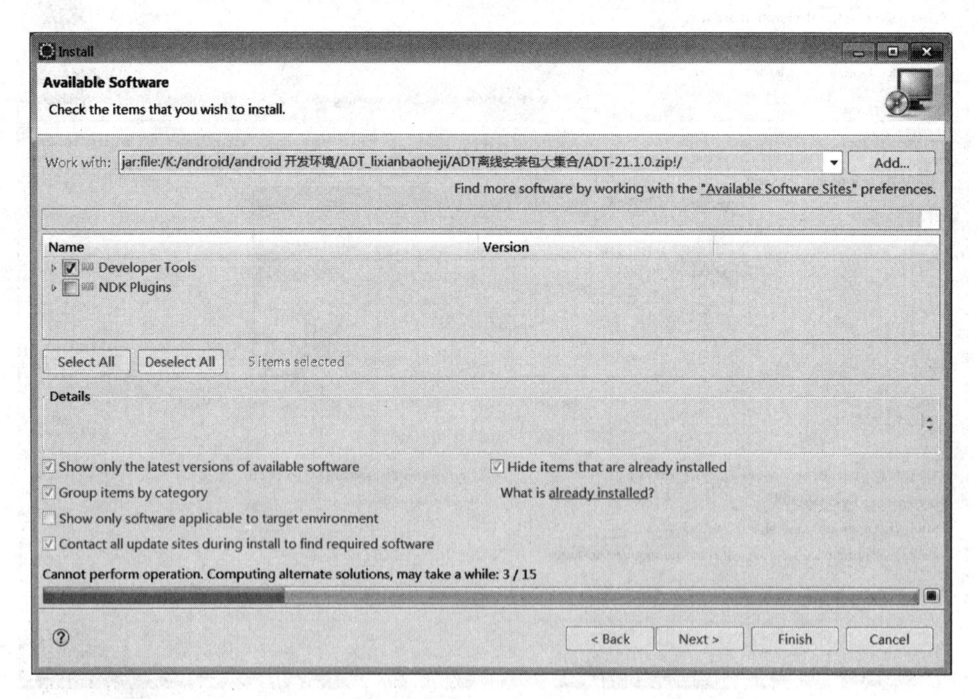

图 2-14　开始安装

4. 配置 SDK

完成 ADT 的安装后，可以单击 Window|Preference，如图 2-16 所示。在 Preferences 窗口中，选择窗口左侧的“Android”选项。在窗口的右侧 SDK Location 中选择或者输入 SDK 文件夹所在的位置，单击“Apply”按钮，然后单击“OK”按钮，完成 SDK 的绑定，如图 2-17 所示。

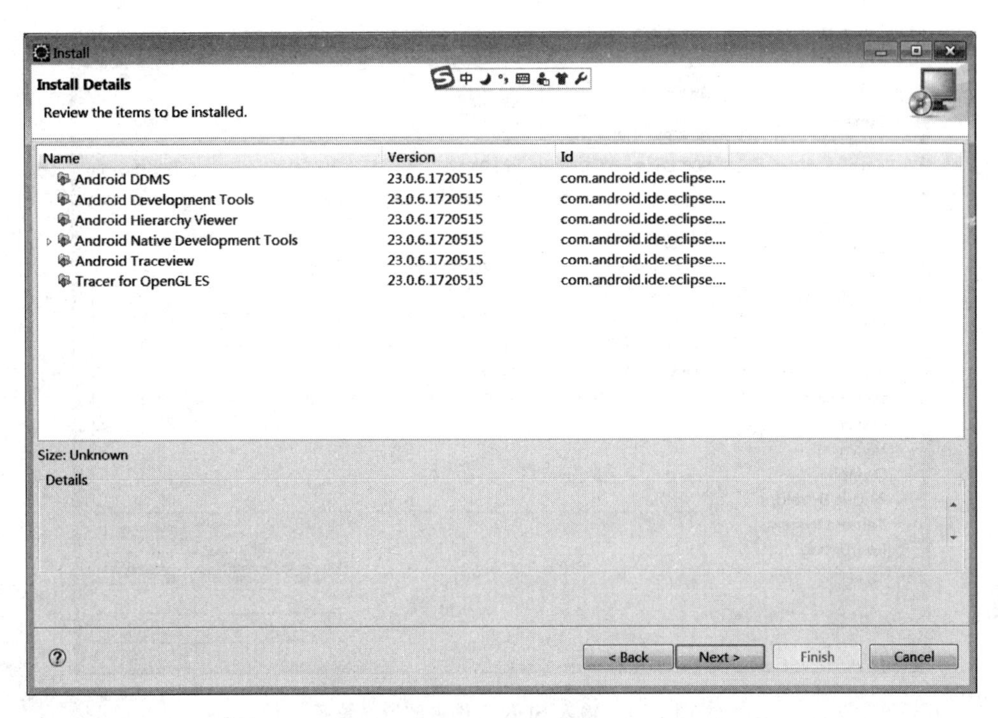

图 2-15　安装完成

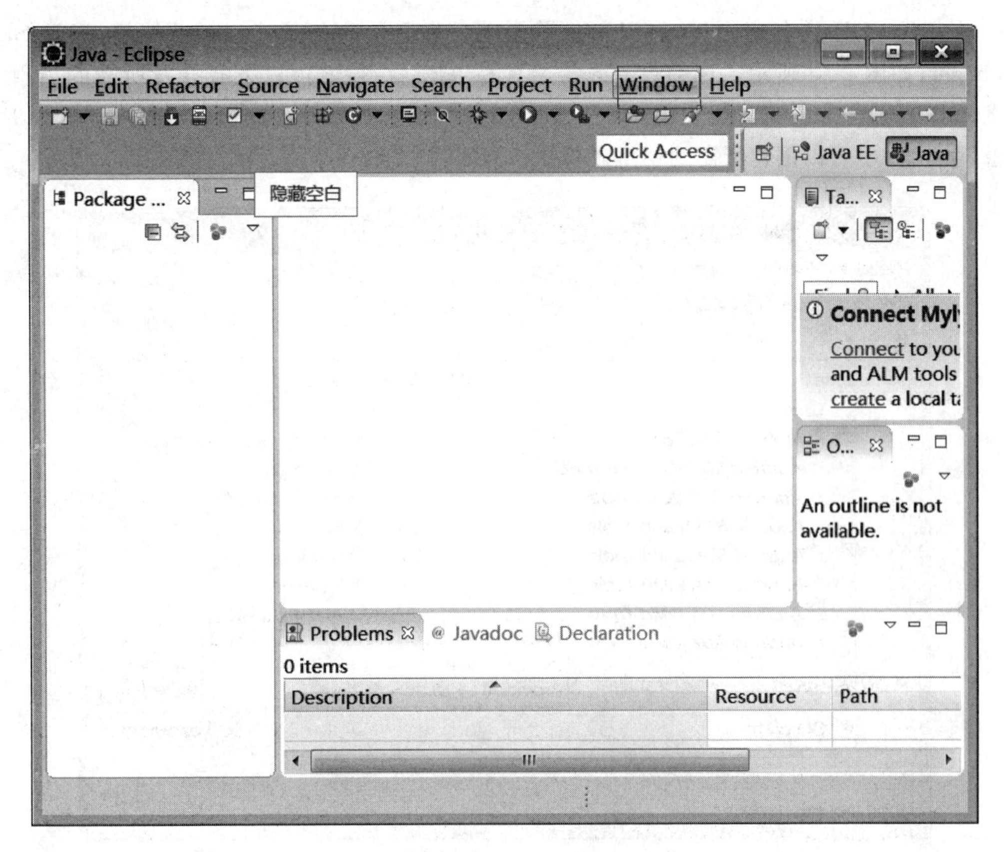

图 2-16　Window 菜单

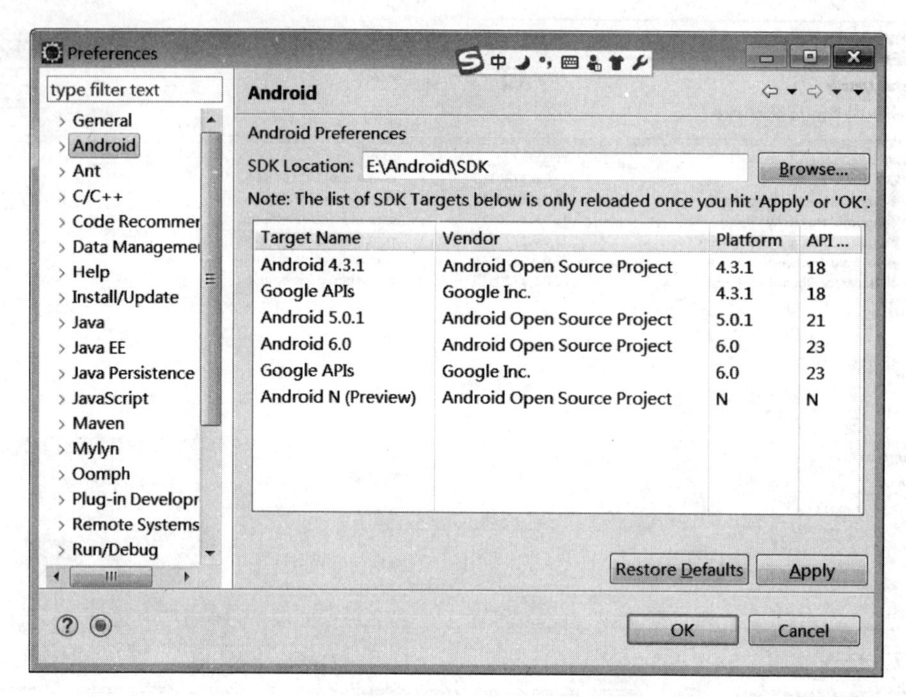

图 2-17 输入 SDK 文件夹的所在位置

在完成 SDK 的绑定之后，可以通过单击工具栏上的“SDK Manager”按钮，启动 SDK Manager，如图 2-18 所示。选中要下载的工具和 API 版本，就可以下载开发所需的 API 和虚拟机工具了。单击工具栏上的“AVD Manager”按钮，启动 AVD Manager，如图 2-19 所示。通过它，就可以创建所需的手机虚拟机了。

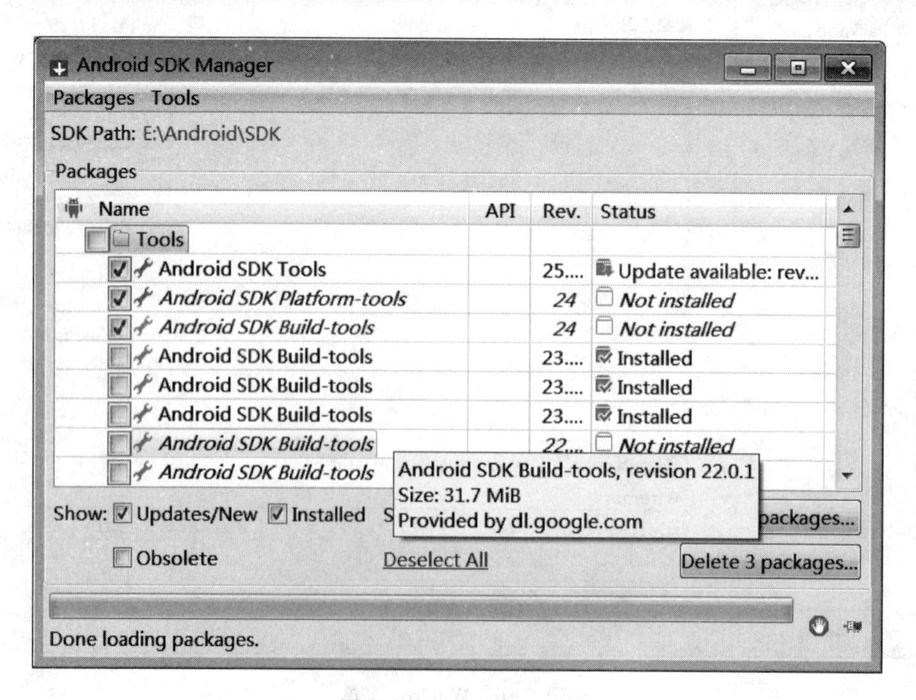

图 2-18 SDK Manager

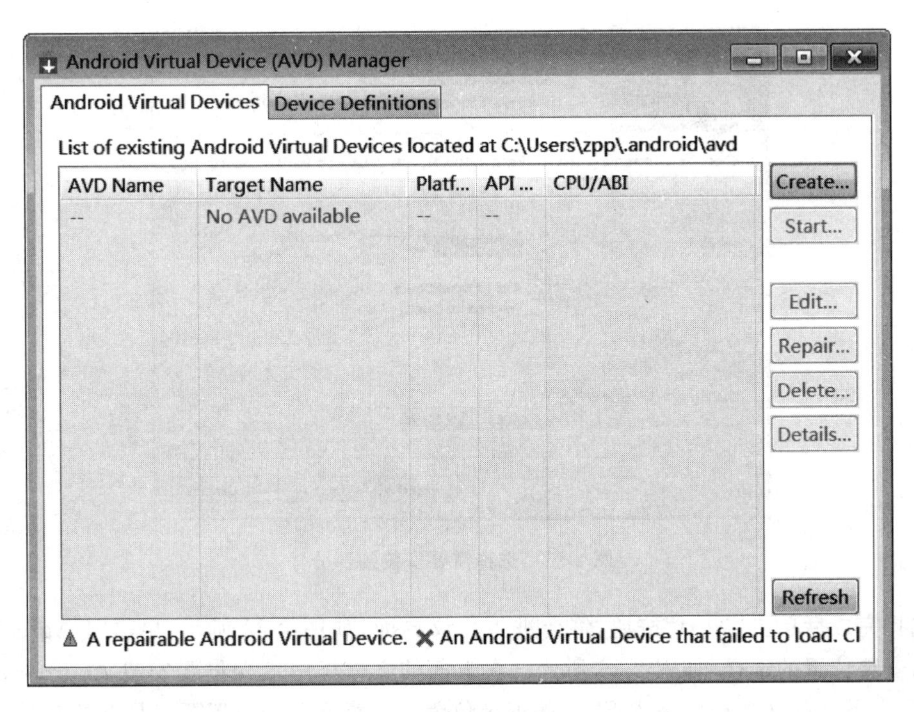

图 2-19 启动 AVD Manager

2.1.3 安装 Android Studio 环境

打开从官网下载的 Android Studio,就会出现安装引导界面,如图 2-20 所示。单击“Next”按钮,进入下一步,如图 2-21 所示。在这一步要选择安装哪些功能。四个选项分别是 Android Studio、Android SDK、Android 虚拟机和 Intel HAX 加速器。如果是初次使用,建议全部安装。如果原本有 SDK,则可以选择安装需要的部分。

图 2-20 安装引导界面

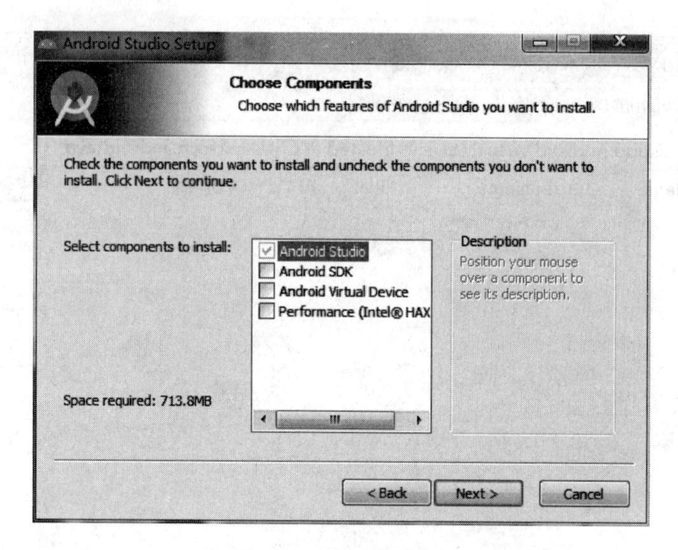

图 2-21 选择需要安装的部分

选择完成后单击“Next”按钮，将会进入协议界面，如图 2-22 所示，单击“I Agree”按钮将进入安装位置的选择，如图 2-23 所示。在上面的文本框中填入将要安装 Android Studio 的位置，在下面的文本框中填入 SDK 的安装位置。注意安装的位置要有足够的空间。

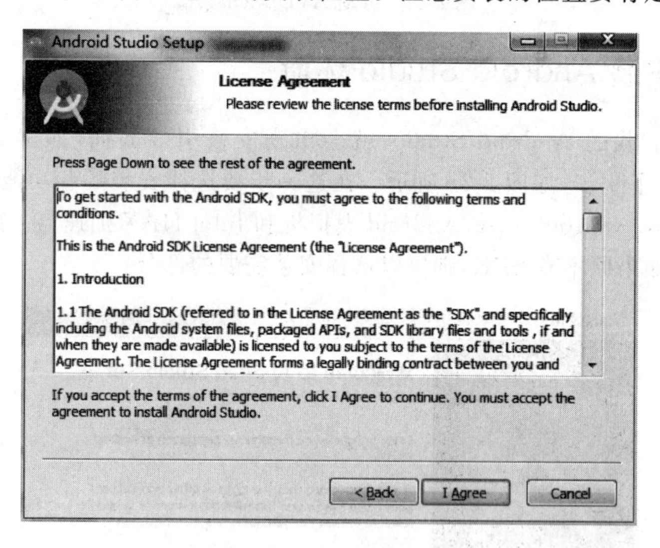

图 2-22 安装协议

选择完成后单击“Next”按钮，这次弹出的窗口是为 HAXM 设置分配多少内存，如图 2-24 所示。单击“Next”按钮进行安装，如图 2-25 所示。安装完成后单击“Next”按钮，在下一个窗口单击“Finish”按钮就完成安装了。

在初次启动 Android Studio 时，还需要选择是否使用之前安装过的 Android Studio 的设置。这个按照情况自己选择就可以了。它还会更新 SDK 工具，稍等一会就可以看到 Android Studio 的欢迎界面，如图 2-26 所示。

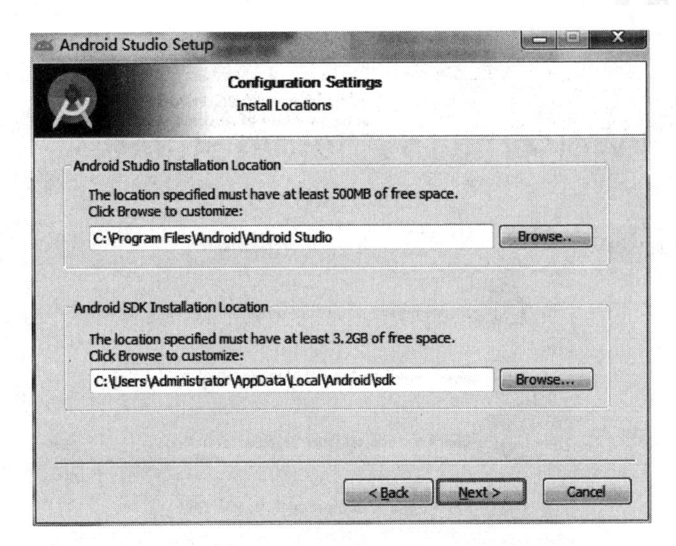

图 2-23　选择安装位置

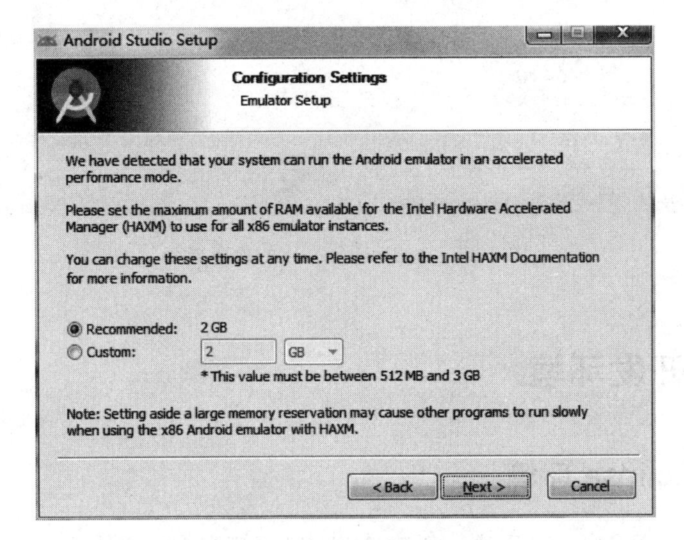

图 2-24　为 HAXM 分配内存

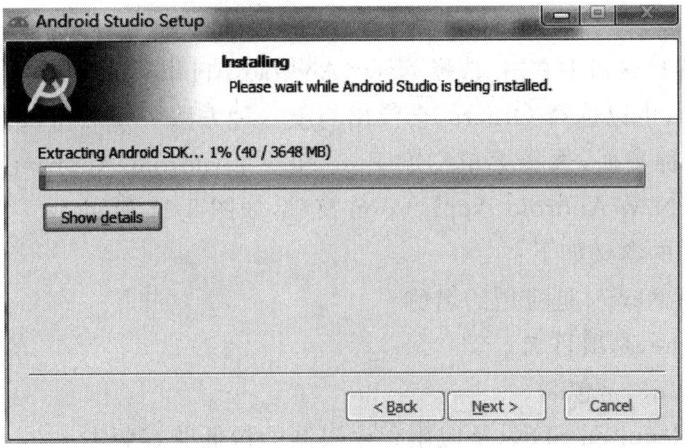

图 2-25　安装进程

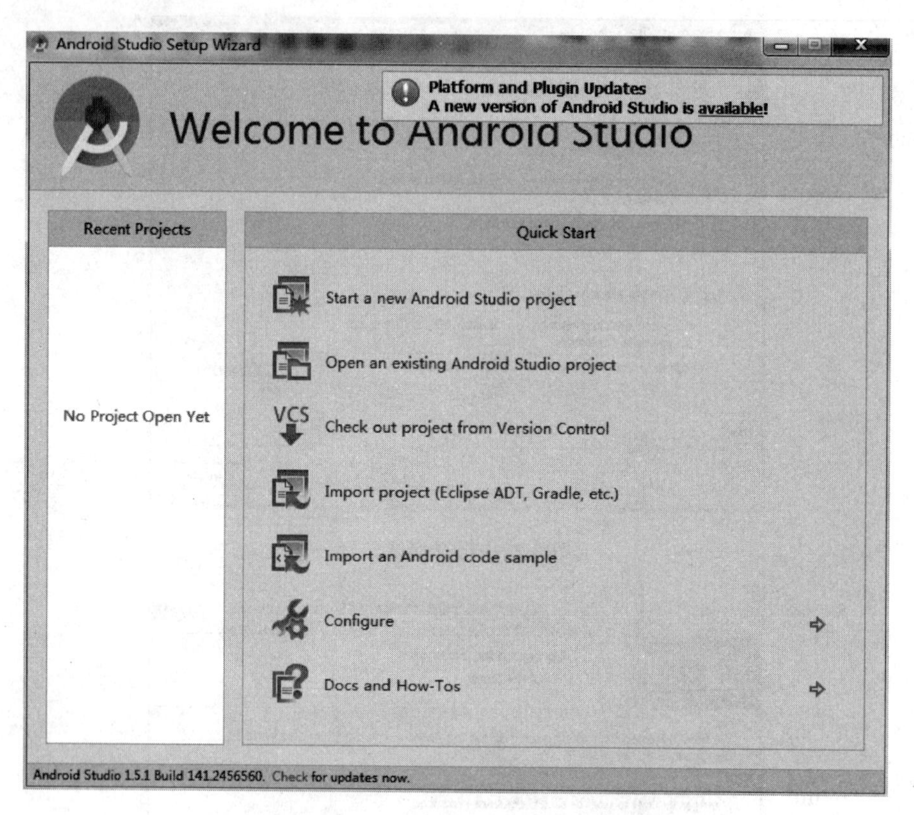

图 2-26 Android Studio 的欢迎界面

2.2 熟悉开发环境

2.2.1 Eclipse 环境

我们先创建一个 Android 程序。新建 Android 程序有多种方法：

第一种，单击快捷栏的“新建”按钮，在弹出的窗口中选择 Android|Android Application Project。

第二种，在工程目录树中右击，选择 New |Android Application Project，如果没有 Android Application Project，可以选择 Others，会弹出与上一种方法一样的窗口。

第三种与第二种类似，选择菜单栏的 File|New |Android Application Project，完成上述操作都可以打开 New Android Application 窗口，如图 2-27 所示。

图 2-27 中各项的含义如下：

- Application Name，是应用的名称。
- Project Name，是项目名。
- Package Name，是包名。
- Minimum Required SDK，是应用程序限制的最低兼容版本。
- Target SDK，是应用的目标版本，即该应用最适合运行的 Android 版本。

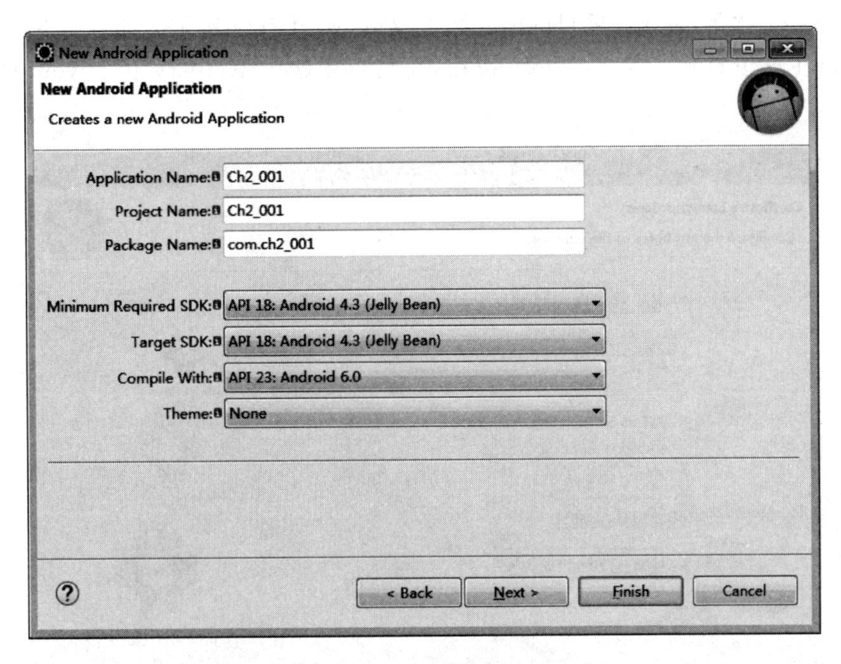

图 2-27 New Android Application 窗口

- Compile With,是编译程序的 SDK 版本。
- Theme,是生成的 UI 所使用的主题。

单击“Next”按钮,进入到创建项目的配置窗口,如图 2-28 所示。这里使用默认设置就可以了。

New Android Application
New Android Application
Configure Project
Create custom launcher icon
Create activity
Mark this project as a library
Create Project in Workspace
Location: E:\workspace\Ch2_001
Browse...
Working sets
Add project to working sets
Working sets:
Select...
< Back
Next >
Finish
Cancel

图 2-28 项目的配置窗口

单击“Next”按钮,进入到项目图标的配置窗口,如图 2-29 所示。如果有自己设计的应用图标可以在这里替换。单击“Next”按钮,在窗口中输入 Activity 和 Layout 的名称,如图 2-30 所示。

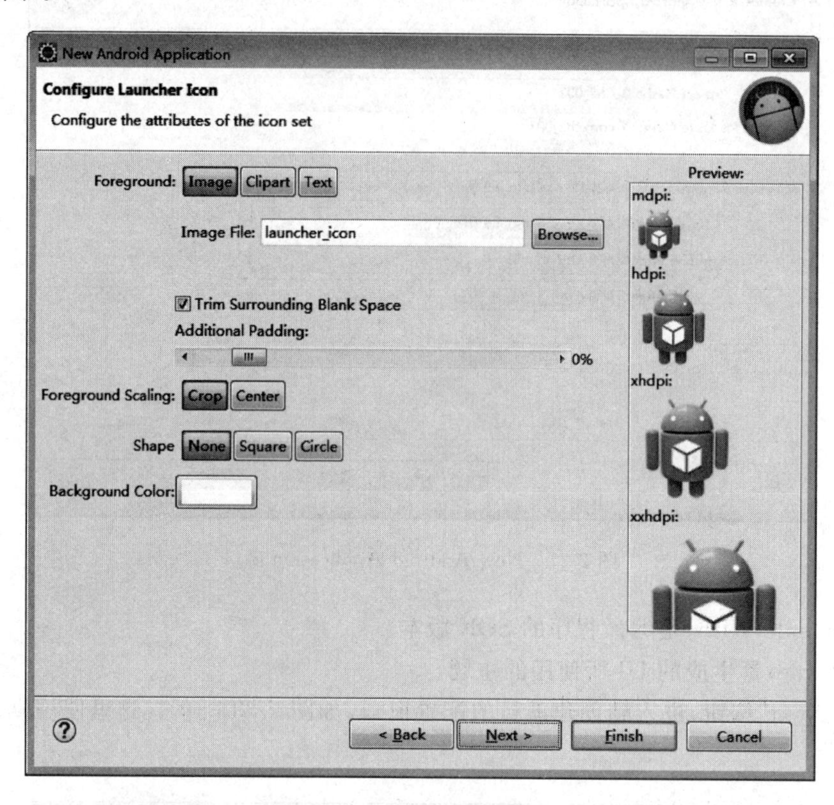

图 2-29 项目图标配置窗口

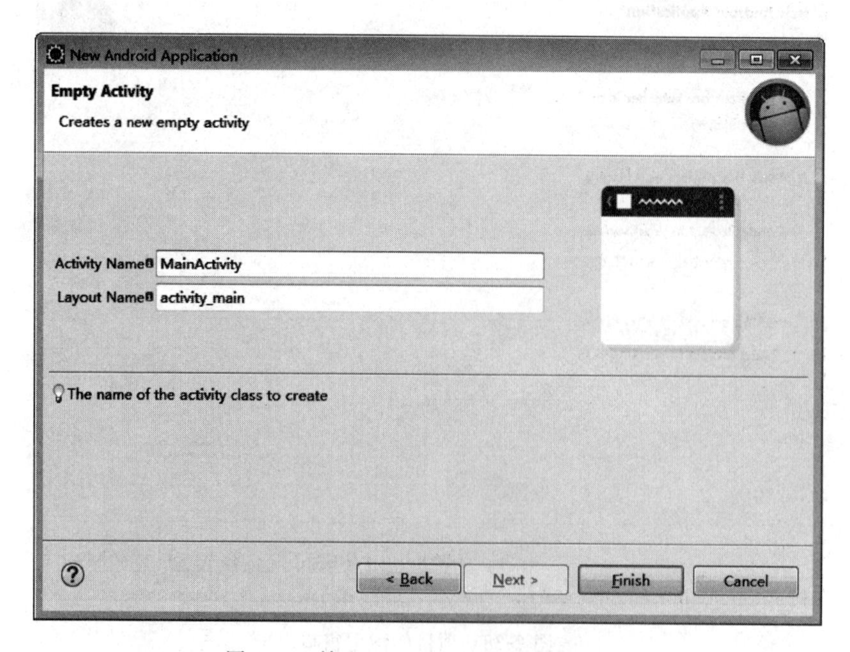

图 2-30 输入 Activity 和 Layout 的名称

单击“Finish”按钮，就完成了一个新的 Android 应用程序的创建。

最后，在菜单栏上单击“运行”，就可以看到实验结果了，如图 2-31 所示。

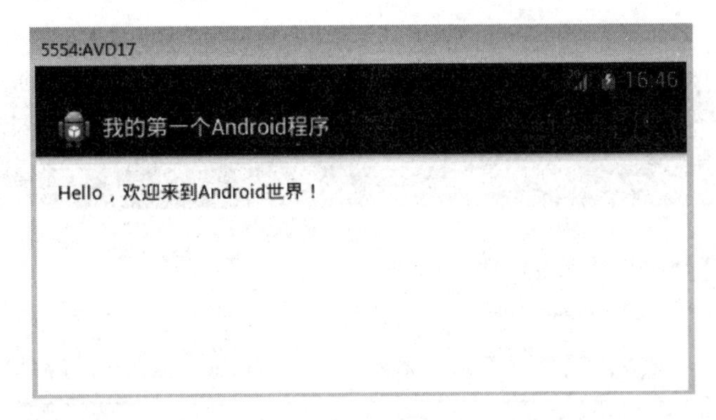

图 2-31　创建的第一个 Android 程序

2.2.2　Android Studio 环境

1. 创建第一个 Android 程序

要在 Android Studio 中新建一个 Android 程序，可以在 Android Studio 欢迎界面选择 Start a new Android Studio project 或者在一个已经打开的项目中单击 File | New | New Project。

在 New Project 界面设定应用的名字、公司域名及项目的存储位置，如图 2-32 所示。

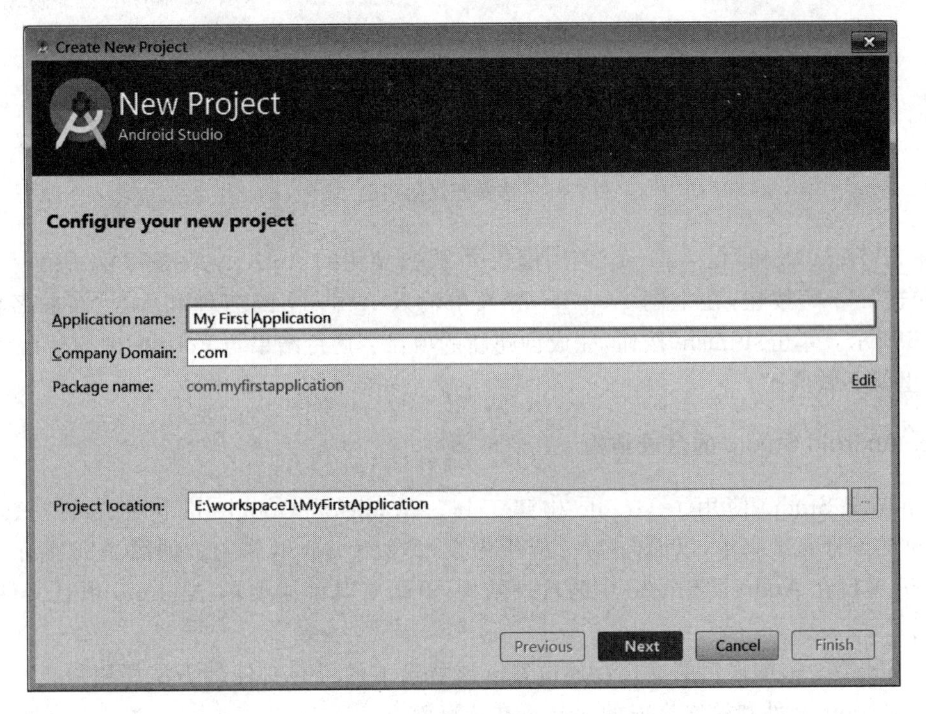

图 2-32　New Project 界面

单击“Next”按钮，打开目标 Android 设备设定界面，选择开发使用的 API，如图 2-33 所示。

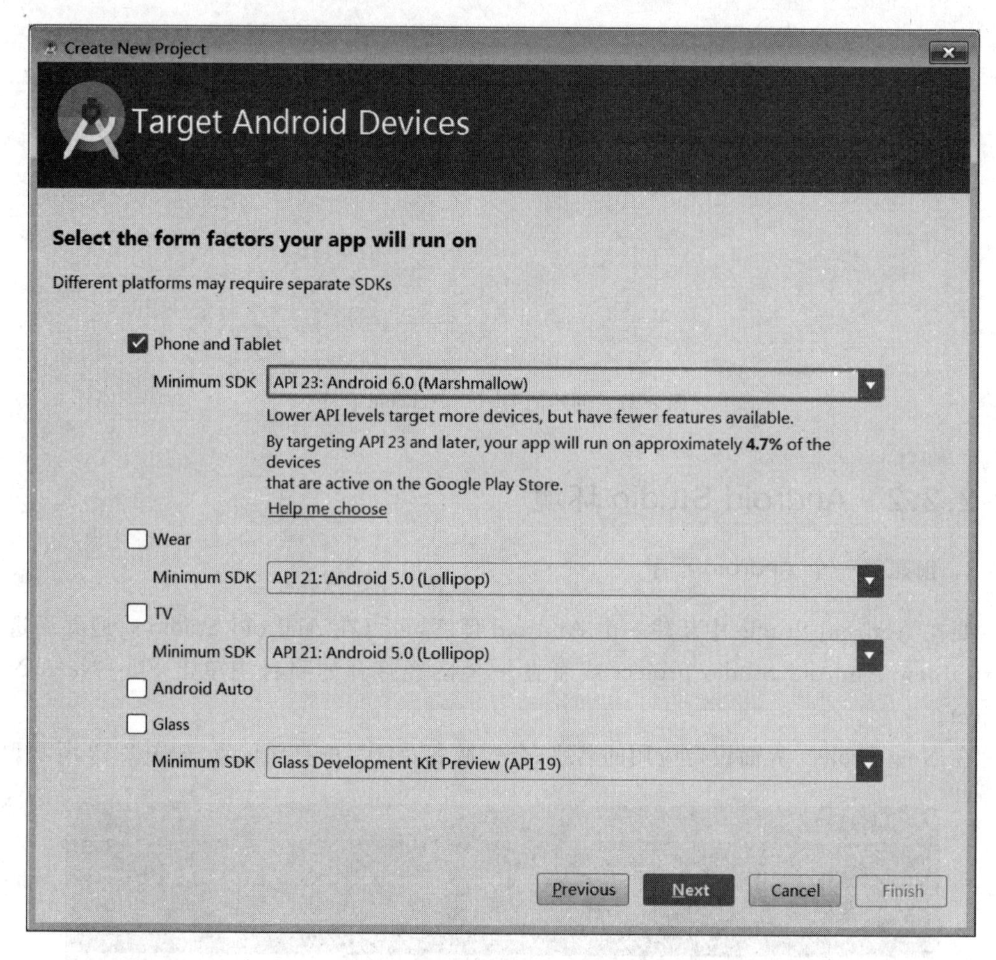

图 2-33　选择开发使用的 API

单击“Next”按钮，在 Activity 应用模板中，选择要用的 Activity，如图 2-34 所示。

单击“Next”按钮，在定制 Activity 界面中输入 Activity 的名称和 Layout 的名称，如图 2-35 所示。单击“Finish”按钮，完成创建新的项目。运行结果与 Eclipse 中的运行结果类似，这里就不做赘述了。

2. Android Studio 的目录结构

Android Studio 使用了 Gradle 构建工具。Gradle 是 Google 推荐使用的一套基于 Groovy 的编译系统脚本，以面向 Java 应用为主，类似于 ant 的作用。如图 2-36 所示，这是 Android 项目在 Android Studio 中的目录结构，下面对其中主要的 Android 项目文件夹进行介绍。

- build/——编译后的文件存放位置。该目录下的所有文件都不需要用户自己创建。R. Java 文件和最终生成的 apk 也在这里面。
- libs/——依赖库所在的位置，存放第三方 jar 包。

Create New Project
Add an Activity to Mobile
Add No Activity
Basic Activity
Empty Activity
Ad
Fullscreen Activity
Google AdMob Ads Activity
Google Maps Activity
Previous
Next
Cancel
Finish

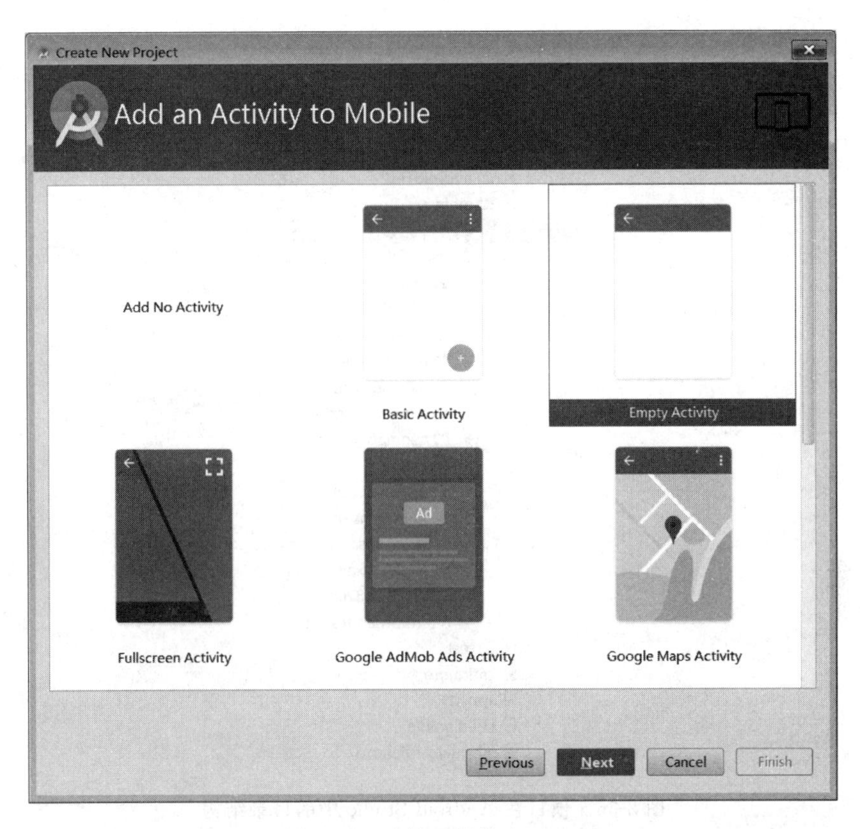

图 2-34 选择要用的 Activity

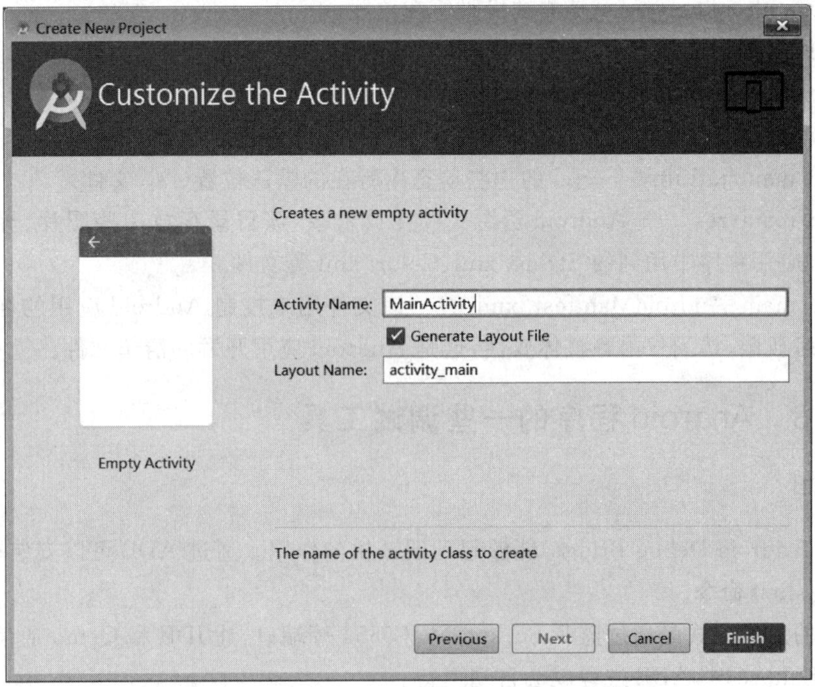

Create New Project
Customize the Activity
Creates a new empty activity
Activity Name: MainActivity
Generate Layout File
Layout Name: activity_main
Empty Activity
The name of the activity class to create
Previous
Next
Cancel
Finish

图 2-35 输入 Activity 名称和 Layout 名称

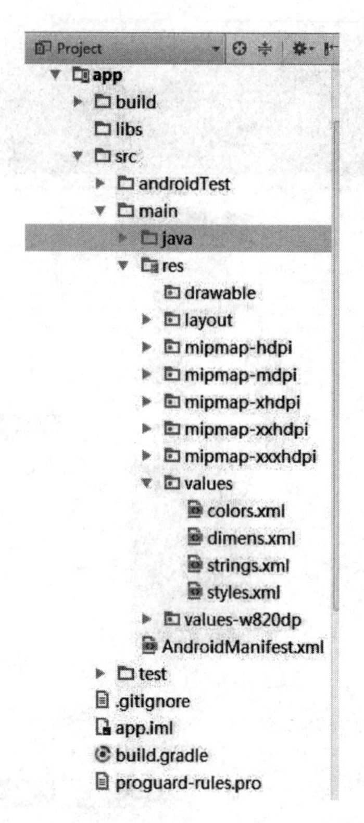

图 2-36　项目在 Android Studio 中的目录结构

- src/——源代码存放的目录，专门存放编写的 Java 源代码。
- src/androidTest/——是测试代码所在位置。
- src/main/——主要代码所在位置。
- src/main/assets/——Android 中附带的一些文件，如视频文件、MP3 等一些媒体文件。
- src/main/Java/——这里是开发者编写代码的文件存放位置。
- src/main/jniLibs/——jni 的一些动态库所在的默认位置(. so 文件)。
- src/main/res/——Android 资源文件存放目录，该目录存放一些图片、界面布局文件、应用程序中用到的 String. xml、Color. xml 等文件。
- src/main/AndroidManifest. xml/——该文件用来控制 Android 应用的名称、图标、网络权限、访问权限等整体属性，这是 Android 应用开发的清单文件。

2.2.3　Android 程序的一些调试工具

1. ADB

ADB(Android Debug Bridge)能够起到调试桥的作用。通过 ADB 可以安装软件、升级系统、运行 shell 命令。

ADB 的工作方式是通过监听 Socket TCP 5554 等端口，让 IDE 和 Qemu 通信。默认情况下运行 Eclipse 时，ADB 进程就会自动运行。

下面是一些简单的 ADB 命令：

(1) 查看设备：adb devices。

(2) 安装软件：adb install，adb install < apk 文件路径>。这个命令可以将指定的 apk 文件安装到设备上。

(3) 卸载软件：adb uninstall <软件名>，adb uninstall -k <软件名>。

(4) 进入设备或模拟器的 shell：adb shell。

(5) 发布端口：adb forward tcp:888 tcp:8001。这里可以设置任意的端口号，设定的端口可以作为主机向模拟器或设备的请求端口。

(6) 从计算机上发送文件到设备：adb push <本地路径> <远程路径>。这里通过 push 命令，可以将本机上的文件或者文件夹复制到手机虚拟机上。

(7) 从设备上下载文件到计算机：adb pull <远程路径> <本地路径>。通过 pull 命令，能够把手机虚拟机上的文件或者文件夹复制到本机。

(8) 查看 bug 报告：adb bugreport。

为了便捷地使用 ADB，可以将存放在 SDK 中的 adb. exe 的目录位置配置到 Path 环境变量中，这样就可以使用命令行窗口来快速操作 ADB 了。

2. DDMS

DDMS(Dalvik Debug Monitor Service)用来在监控及调试 Android 系统中 Dalvik 虚拟机的服务。DDMS 在 IDE、模拟器和真机之间起到桥梁的作用。它能通过 ADB 建立调试桥，进而将指令发送到调试的终端。

它可以提供多种服务。这些服务包括：查看正在运行的线程以及堆信息、截屏、LogCat、广播状态信息、模拟电话呼叫、接收短信、设定虚拟地理坐标等。

DDMS 被集成在了 Eclipse 中，在 SDK 下 tools 目录下也有 DDMS。

在 Eclipse 中打开 DDMS 的方法是：在菜单栏中单击 Window | Open Perspective | Other，在弹出的窗口中双击 DDMS 就可以启动 DDMS。

在 Android Studio 中打开 DDMS 的方法是：在 Android Studio 的菜单栏单击 Tools | Android | Android Device Monitor。在 SDK 的 tools 路径下存放了 ddms. bat，直接双击 ddms. bat 就可以启动 DDMS。

在启动的 DDMS 中，左侧窗格中的 Devices 中显示所有与 DDMS 连接的模拟器的详细信息，例如每个模拟器正在执行的 APP 的进程，以及模拟器相对应的进程与调试器连接的端口号。在右侧窗格中可以看到线程信息(Threads)、内存分配情况、网络情况、文件资源管理器(File Explore)、仿真控制器(Emulator Control)等。它们都是以标签的形式存在的，可以随意拖动或者关闭，所以不同的人的排列顺序或显示情况可能会出现不同。下方的窗口是 LogCat，主要用于输出一些模拟器的信息。

下面介绍一下文件资源管理器(File Explore)和仿真控制器(Emulator Control)的一些功能。

Emulator Control 可以实现对模拟器的控制，例如接听电话、模拟接收短信和发送用于测试 GPS 功能的虚拟位置坐标等。

在 Emulator Control 中：

- Telephony Status：可以选择模拟通话质量以及信号连接模式。

- Telephony Actions：可以模拟接听电话和发送短信。
- Location Control：可以模拟地理坐标。

使用DDMS模拟发送短信的操作过程如下：在Emulator Control|Telephony Actions中输入要发送的内容，单击“发送”按钮，然后在Android模拟器中打开短信，即可看到刚刚发送的短信。

File Exporter是文件浏览器，通过它可以查看Android模拟器中的文件，并且可以很方便地导入/导出文件。

2.3 不同环境之间的转换

Android Studio是一款由Google官方开发、指定的Android开发工具。然而也有许多使用Eclipse开发的项目。那么应该如何将Eclipse下的工程项目导入到Android Studio中呢？将Eclipse下的工程项目导入到Android Studio中，生成Android Studio项目的步骤如下：

(1) 在Eclipse的项目栏里右击File|Export。

(2) 在弹出的窗口中选择Generate Gradle build files，然后单击“Next”按钮。

(3) 选中要导出的工程，单击“Next”按钮。

(4) 提示选择将要导出的Gradle文件的存放位置，选择好项目的存放位置后单击“Finish”按钮。

(5) 打开Android Studio，单击菜单栏File|Import Project。

(6) 在弹出的对话框中选择刚刚导出的工程，然后单击“OK”按钮，完成操作。

2.4 不同环境之间的比较

既然介绍了这两种不同的开发工具，那么使用时如何选择呢？下面对比一下它们的各自优势。

1. Android Studio可以更方便地构建程序界面

在Eclipse中构建的应用程序界面，其效果与真机上的差别大，反应速度也不理想。相反，在Android Studio中构建的界面，显示效果非常清晰，且易于迅速修改。

2. Android Studio拥有更详细的信息

Android Studio拥有十分详细的信息。这些信息包括几乎所有在项目中可能遇到的问题。例如编写、设计、开发、打包、构建过程中出现的错误，都可以在控制台上显示出来。这种设计有利于在开发过程中随时发现和定位问题。相反，Eclipse中的信息则相对要少得多，使得许多问题不能及时发现，增大了开发的难度。

3. Android Studio拥有更详细的编辑历史

在Android Studio使用过程中，修改代码和布局文件或者删除文件这些操作会生成非常细致的记录。这个过程中的每一个操作都有记录，并且每一个操作都能够撤销。与之不同，在 Eclipse 中执行了删除文件的操作后，它的编辑记录就会被清空，因而无法完成回滚操作。

4. Android Studio 能够预览资源文件

Android Studio 可以在开发的过程中实时地预览资源文件的内容，而 Eclipse 就没有这项功能。

5. Eclipse 可以更简单地创建项目

在 Eclipse 中创建一个项目其过程十分方便、快捷，而在 Android Studio 创建项目的过程中，可能遇到各种 Gradle 构建的问题。

6. Eclipse 中的项目体积比较小

在 Eclipse 中创建的项目体积比 Android Studio 中的项目体积更小。在 Android Studio 的项目中，需要各种各样的清单文件，这些文件包含了各种工具自身的历史文件，还有 Gradle 的构建文件等，这些文件在 Eclipse 的项目中可能并不存在。所以 Android Studio 中项目的体积就比 Eclipse 大得多。

在 Eclipse 中创建好的清单文件无须更新，而 Android Studio 则会频繁地更新各种文件。

7. Eclipse 中管理多项目更方便

在 Eclipse 中多个项目可以放在一个工作空间，可以非常方便地进行多项目的管理。而 Android Studio 中每打开一个项目都需要启动一次程序。

这两种开发工具各有优劣，但是 Android Studio 是谷歌官方开发的并且在不断地完善，明显会拥有更好的前景。

2.5 本章小结

本章介绍了 Android 开发环境的搭建和设置，讲解了两种搭建方法。尽管 Android Studio 是官方推荐的开发环境，但由于 Eclipse 过去使用者众多，本章同样简单介绍了它的安装和使用，并介绍了项目迁移的方法。

2.6 练习题

一、填空题

1. ADB 的常见指令中用于列出设备的是________。
2. JDK 的安装完成后还需要配置________。
3. Android 在 Eclipse 环境下开发需要________和________的支持。

二、选择题

1. 以下目录用来存放源文件的是(　　)。

A. src 目录　　B. libs 目录　　C. assets 目录　　D. res 目录

2. 用 Android Studio 开发 Android 程序不需要(　　)。

A. ADT　　B. SDK　　C. JDK　　D. AVD

三、简答题

1. 简单描述 Android Studio 环境搭建的过程。
2. 试述如何导入新项目。
3. 试述如何新建一个项目。
4. 将一个 Eclipse 项目导入 Android Studio。

第3章 Activity及其生命周期

本章重点

- Activity的创建
- Activity的生命周期
- JUnit单元测试
- 资源的引用

一个Android编写的应用是由活动(Activity)、意图(Intent)、广播接收器(Broadcast Receiver)、内容提供器(Content Provider)这四大部分组成的。

其中,Activity(活动)是Android一个APP中最基本的构成部分之一。Activity类似于Java SE中的窗体,一个Activity对象代表一个屏幕,用来创建显示窗口。在这四个组件中,Activity是用户唯一可以看得到的组件,它通过一个Activity栈来进行管理。同时,用户可以在Activity当中添加TextView、Button、Checkbox等组件与Activity进行交互。

一般情况下,一个Android应用是由多个Activity组成的。不同的Activity之间可以相互跳转,这个过程需要在资源清单文件AndroidManifest.xml中添加权限。

3.1 Activity的创建

在一个Android应用中可以有一个或者多个Activity。在创建一个Android项目的时候,系统已经生成了一个Activity。那么如何自己创建一个Activity呢?

创建Activity的步骤如下:

(1) 在src目录中建立一个类,这个类继承自Activity或其子类。在res/layout目录中建立一个后缀名为.xml的布局文件。

(2) 在新建的类文件中重写onCreate()方法,加载指定的布局文件。

(3) 改写清单文件AndroidManifest.xml。

(4) 右击"包名",如图3-1所示,选择菜单New|Class。

(5) 在弹出的"New Java Class"窗口中输入要建立的类名和继承自哪个类,如图3-2所示。

(6) 在"Name"文本框中输入名称,本例是Activity2。单击"Superclass"后的"Browse"按钮,设置为继承自"android.app.Activity"类,然后单击"Finish"按钮完成创建。

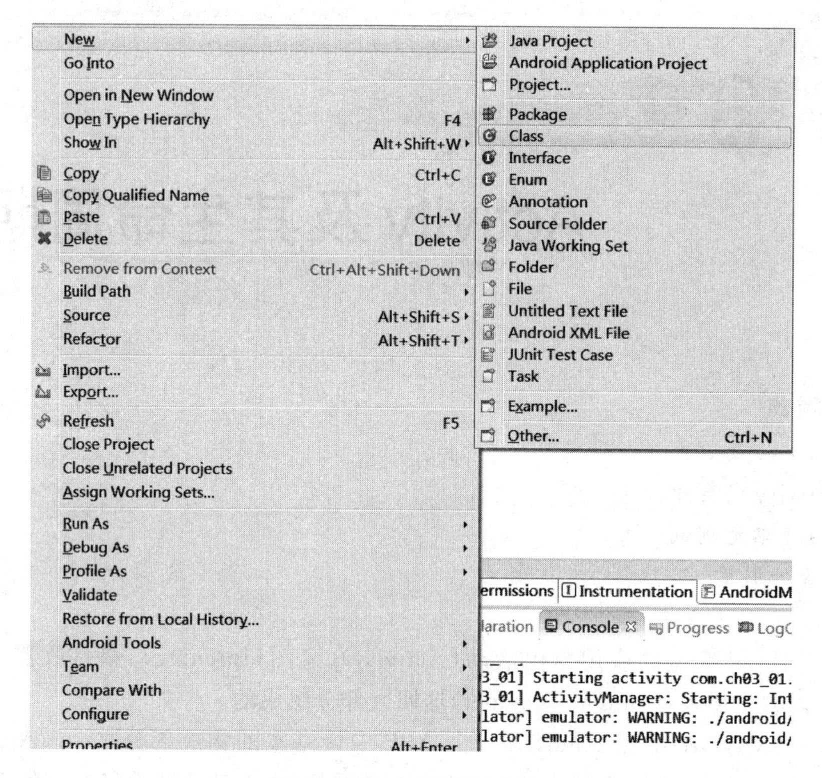

图 3-1 “包名”的右键菜单

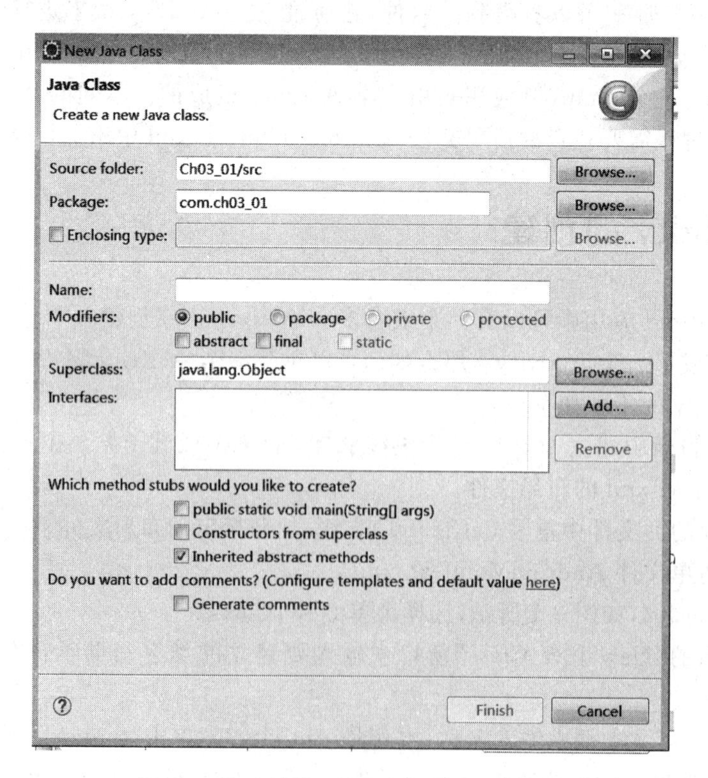

图 3-2 输入要建立的类名和继承自哪个类

创建完成后，打开 Activity2. java 文件能够看到里面的内容，如图 3-3 所示。

```
1 package com.ch03_01;
2
3 import android.app.Activity;
4
5 public class Activity2 extends Activity {
6
7 }
8
```

图 3-3 Activity2. java 文件内容

（7）右击 res/layout 文件夹，如图 3-4 所示，选择菜单 New|Android XML File，在弹出的“New Android XML File”窗口中，输入布局文件的文件名，选择要建立的布局类型，如图 3-5 所示。最后单击“Finish”按钮完成创建。

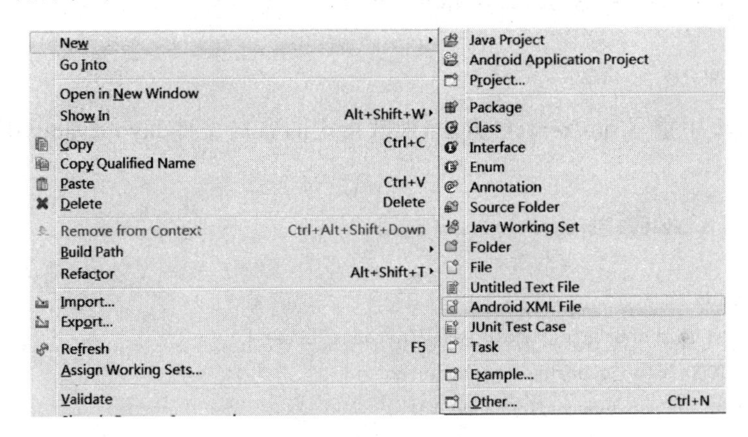

图 3-4 新建布局文件

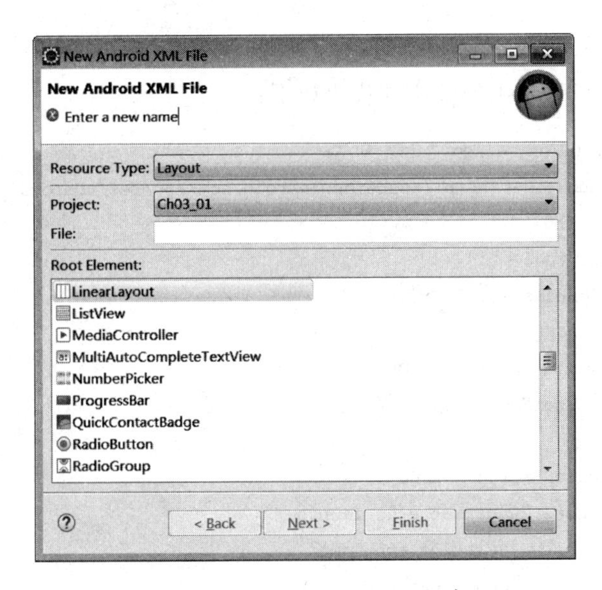

图 3-5 New Android XML File 窗口

本实验中建立的布局文件 layout_activity2. xml 代码如下：

```
<?xml version = "1.0" encoding = "utf - 8"?>
<LinearLayout xmlns:android = "http://schemas.android.com/apk/res/android"
    android:layout_width = "match_parent"
    android:layout_height = "match_parent"
    android:orientation = "vertical" >

    <TextView
        android:id = "@ + id/textView1"
        android:layout_width = "wrap_content"
        android:layout_height = "wrap_content"
        android:textSize = "31dp"
        android:textColor = "#00ff00"
        android:text = "这是第二个 Activity" />

</LinearLayout>
```

在 Activity2 中重写 onCreate()方法，加载指定的布局文件 layout_activity2. xml，其代码如下：

```
public class Activity2 extends Activity {

    @Override
    protected void onCreate(Bundle savedInstanceState) {
        // TODO Auto - generated method stub
        super.onCreate(savedInstanceState);
        setContentView(R.layout.layout_activity2);
    }
}
```

原本的 AndroidManifest. xml 文件代码如下：

```
<?xml version = "1.0" encoding = "utf - 8"?>
<manifest xmlns:android = "http://schemas.android.com/apk/res/android"
    package = "com.ch03_01"
    android:versionCode = "1"
    android:versionName = "1.0" >
    <uses - sdk
        android:minSdkVersion = "18"
        android:targetSdkVersion = "21" />
    <application
        android:allowBackup = "true"
        android:icon = "@drawable/ic_launcher"
        android:label = "@string/app_name"
        android:theme = "@style/AppTheme" >
        <activity
```

```
            android:name = ".MainActivity"
            android:label = "@string/app_name" >
            <intent - filter>
                <action android:name = "android.intent.action.MAIN" />

                <category android:name = "android.intent.category.LAUNCHER" />
            </intent - filter>
        </activity>
    </application>
</manifest>
```

运行程序,运行结果如图 3-6 所示。

将 AndroidManifest. xml 中< Activity >标签下的 android:name=". MainActivity"改为 android:name=". Activity2",其运行结果如图 3-7 所示。

这种操作在程序运行的过程中只调用了一个 Activity。关于多个 Activity 切换的内容,将在 Intent 的相关章节进行介绍。

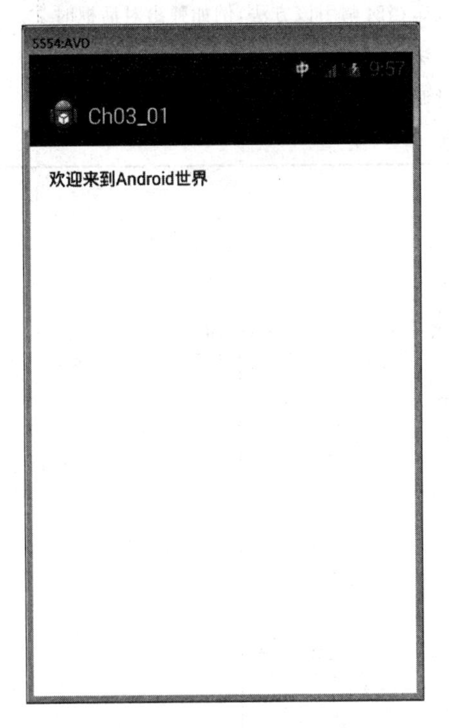

图 3-6　运行结果

图 3-7　改动后的运行结果

3.2　Activity 的生命周期

3.2.1　Activity 生命周期的概念

Activity 的生命周期,就是一个 Activity 从创建到销毁的过程。生命周期的过程中共有四种状态：

（1）激活或者运行状态：此时的 Activity 运行在屏幕的前端；

（2）暂停状态：此时的 Activity 失去了焦点，但是对用户仍然可见；

（3）停止状态：此时的 Activity 被其他的 Activity 覆盖；

（4）终止状态：此时的 Activity 将被销毁，释放内存资源。

Activity 的生命周期的四种状态，可通过表 3-1 中的 8 种方法实现，其状态转换如图 3-8 所示。

表 3-1 Activity 类中的方法

方　法	功能描述
onCreate()	Activity 初次创建时调用该方法，一些组件的声明和调用就写在该方法中
onStart()	当 Activity 可见的时候调用该方法
onRestart()	当 Activity 再次可见时调用该方法
onResume()	当 Activity 获取焦点调用此方法
onFreeze()	当 Activity 被暂停而没有完全被遮挡时调用该方法，例如弹出对话框时
onPause()	当 Activity 失去焦点时调用此方法
onStop()	当 Activity 被完全遮挡不可见时调用此方法
onDestroy()	在 Activity 被销毁时调用此方法

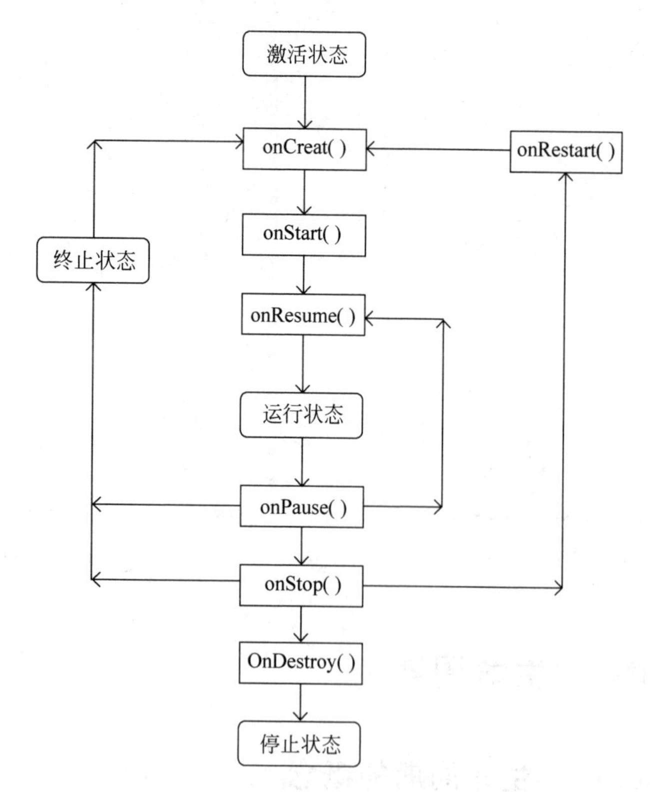

图 3-8 Activity 的生命周期

3.2.2 Activity 生命周期的案例

在 MainActivity. java 中重写 onCreate(),onStart(),onRestart(),onResume(),onFreeze(),onPause(),onStop(),onDestroy()方法,调用 Log 类中的 log. d()输出日志。日志的 tag 设置为 MainActivity,输出信息设置为对应的方法名。具体代码如下:

```
package com.ch03_01;
import android.app.Activity;
import android.os.Bundle;
import android.util.Log;

public class MainActivity extends Activity {

    @Override
    protected void onCreate(Bundle savedInstanceState) {
        super.onCreate(savedInstanceState);
        setContentView(R.layout.activity_main);
        Log.d("MainActivity", "onCreate()");
    }
    // 当这个 Activity 变成用户可见时调用的方法
    protected void onStart() {
        super.onStart();
        Log.d("MainActivity", "onStart()");
    }
    protected void onRestart() {
        super.onRestart();
        Log.d("MainActivity", "onRestart()");
    }
    // 当 Activity 获取到焦点时调用的方法
    protected void onResume() {
        super.onResume();
        Log.d("MainActivity", "onResume()");
    }
    // 当 Activity 失去焦点时调用的方法
    protected void onPause() {
        super.onPause();
        Log.d("MainActivity", "onPause()");
    }
    // 当 Activity 对用户不可见时调用的方法
    protected void onStop() {
        super.onStop();
        Log.d("MainActivity", "onStop()");
    }
    // 当 Activity 被销毁时调用的方法
    protected void onDestroy() {
        super.onDestroy();
        Log.d("MainActivity", "onDestroy()");
    }
}
```

运行结果如图 3-9 所示。

使用 Eclipse 中的 LogCat 捕捉关于此 Activity 生命周期的结果,如图 3-10 所示。

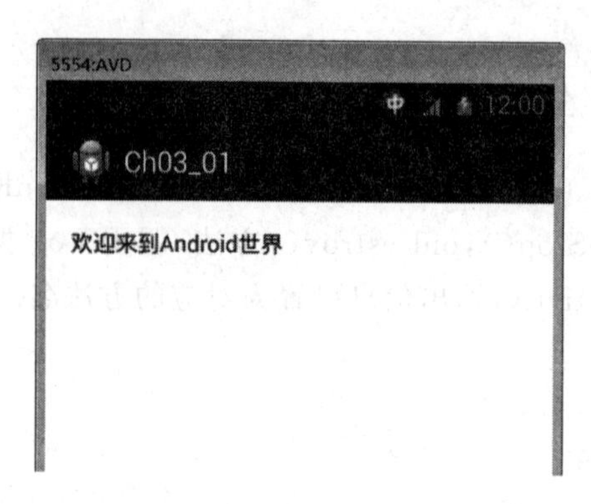

图 3-9 运行结果

Problems Javadoc Declaration Console Progress LogCat

Saved Filters
All messages (n
MyActivity (3)

Search for messages. Accepts Java regexes. Prefix with pid:, app:, ta verbose

L...	Time	PID	TID	Application	Tag	Text
I	07-09 03:17:...	1205	1421	system_process	Activity...	START u0 {a
						nt.category
						Activity} f
W	07-09 03:17:...	1205	1421	system_process	WindowMa...	Screenshot
						r 21005
I	07-09 03:17:...	2286	2286	com.ch03_01	MainActi...	onCreate()
I	07-09 03:17:...	2286	2286	com.ch03_01	MainActi...	onStart()
I	07-09 03:17:...	2286	2286	com.ch03_01	MainActi...	onResume()
D	07-09 03:17:...	2314	2314		AndroidR...	Shutting do
D	07-09 03:17:...	2314	2321		dalvikvm	Debugger ha
I	07-09 03:17:...	1205	1221	system_process	Activity...	Displayed c

图 3-10 捕捉关于此 Activity 生命周期的结果

可以看到，LogCat 显示的结果十分繁杂。为了能够方便地找到需要的信息，需要先设置 LogCat。

如图 3-11 所示，LogCat 窗口的右侧可以选择显示哪一个级别的日志信息。级别分类和调用方法如表 3-2 所示。如果使用默认设置，只能大致找到不同级别的信息。如果想精确地找到所需日志，可以设置过滤器。

Saved Filters
All messages (n
MyActivity (3)

Search for messages. Accepts Java regexes. Prefix with pid:, app:, ta verbose

verbose
debug
info
warn
error
assert

L...	Time	PID	TID	Application	Tag	Text
D	07-09 03:17:...	2302	2302		Andro	>>> Andr
						it <<<<<
D	07-09 03:17:...	2302	2302		Andro	ckJNI is
D	07-09 03:17:...	2302	2302		dalvi	ying to l
D	07-09 03:17:...	2302	2302		dalvikvm	Added share
D	07-09 03:17:...	2302	2302		dalvikvm	Trying to l
D	07-09 03:17:...	2302	2302		dalvikvm	Added share
D	07-09 03:17:...	2302	2302		AndroidR...	Calling mai
D	07-09 03:17:...	2302	2302		AndroidR...	Shutting do
D	07-09 03:17:...	2302	2309		dalvikvm	Debugger ha
D	07-09 03:17:...	2314	2314		AndroidR...	>>>>>> Andr

图 3-11 LogCat 日志信息级别

表 3-2 Log类常用的静态方法

调用方法	级别分类	功能说明	显示的颜色
log. v()	verbose	所有信息,最低级别	黑色
log. d()	debug	调试信息	蓝色
log. i()	info	一般信息	绿色
log. w()	warn	警告信息	橙色
log. e()	error	错误信息	红色
log. wtf()	assert	严重异常信息	红色

单击图 3-11 中左侧加号可以创建过滤器,其设置内容如图 3-12 所示。

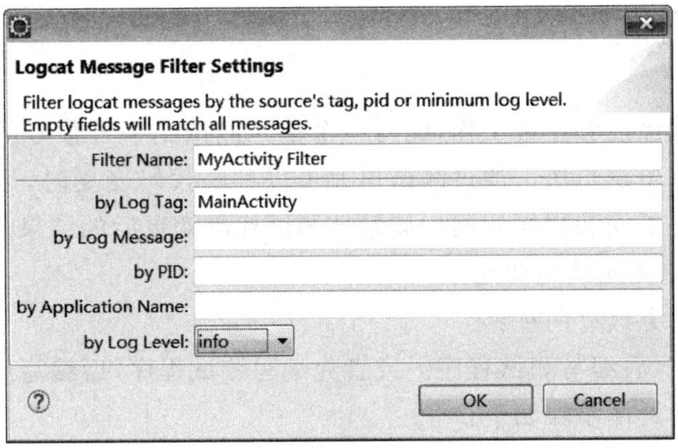

图 3-12 LogCat 过滤器设置

可以看到,过滤器可以设置不同的方法进行过滤。分别是:

- by Log Tag:自定义的标签。
- by Log Message:输出内容。
- by PID:进程的 ID。
- by Application Name:应用的名称。
- by Log Level:日志的级别。

当程序启动时,如果选择 by Log Tag 过滤,则 LogCat 中显示的日志信息如下:

```
07 - 07 02:32:46.935: I/MainActivity(2060): onCreate()
07 - 07 02:32:46.939: I/MainActivity(2060): onStart()
07 - 07 02:32:46.940: I/MainActivity(2060): onResume()
```

由此可知当一个 Activity 启动时需要依次调用的方法。

当按下"返回"按钮时,LogCat 中显示的日志信息如下:

```
07 - 07 02:34:46.527: I/MainActivity(2060): onPause()
07 - 07 02:34:47.409: I/MainActivity(2060): onStop()
07 - 07 02:34:47.409: I/MainActivity(2060): onDestroy()
```

由此可知,当一个 Activity 关闭时需要依次调用 onPause()、onStop()和 onDestroy()方法。

在启动状态中,收到短信并查看短信时,显示的日志信息如下:

```
07 - 07 02:35:52.438: I/MainActivity(2060): onPause()
07 - 07 02:35:52.451: I/MainActivity(2060): onStop()
```

由此可知，当一个 Activity 挂起，需要依次调用 onPause()、onStop()方法。

关闭短信并返回应用时，日志信息如下：

```
07 - 07 02:36:16.153: I/MainActivity(2060): onRestart()
07 - 07 02:36:16.154: I/MainActivity(2060): onStart()
07 - 07 02:36:16.154: I/MainActivity(2060): onResume()
```

即当一个 Activity 挂起返回时，需要依次调用 onRestart()、onStart()和 onResume()方法。

3.3 JUnit 测试

为了确认应用程序能够正常工作，编写一个单元测试是一种很好的方法。单元测试被用来测试程序中的一个逻辑块。通过测试单元可以判断代码变更时，运行的结果是否符合预期。这种测试避免了安装程序后逐一场景地测试代码变更后的结果。基于 JUnit 测试框架，Android 提供了单元测试的功能。

编写测试单元的方式有两种：

先编写应用程序，后编写测试程序。或者先编写测试程序，后编写应用程序。

创建单元测试的具体步骤如下：

(1) 创建一个 Android Test Project 项目，方法：File|New|Project|Android Test Project。

(2) 配置 AndroidManifest. mxl 文件，具体代码如下：

```
<?xml version = "1.0" encoding = "utf - 8"?>
<manifest xmlns:android = "http://schemas.android.com/apk/res/android"
    package = "com.zyz"
    android:versionCode = "1"
    android:versionName = "1.0" >

    <uses - sdk
        android:minSdkVersion = "18"
        android:targetSdkVersion = "21" />
    <instrumentation
    android:targetPackage = "com.zyz" android:name = "android.test.InstrumentationTestRunner">
    </instrumentation>
    <application
        android:allowBackup = "true"
        android:icon = "@drawable/ic_launcher"
        android:label = "@string/app_name"
        android:theme = "@style/AppTheme" >
        <uses - library android:name = "android.test.runner"/>
        <activity
            android:name = "com.zyz.MainActivity"
            android:label = "@string/app_name" >
            <intent - filter>
```

```
                <action android:name = "android.intent.action.MAIN" />

                <category android:name = "android.intent.category.LAUNCHER" />
            </intent - filter>
        </activity>
    </application>

</manifest>
```

说明：在清单文件中添加指令集< instrumentation >，在< application >节点下配置< uses-library >。注意配置的包名 android:targetPackage=""，必须与应用的包名一致。

(3) 创建 JUnit 测试类。创建的 JUnit 类继承自 AndroidTestCast 类，在该类中创建一个方法用于测试。创建的测试方法必须把异常抛出，这样才能得到测试结果。方法的代码如下所示：

```
public void Add() throws Exception
{
int x = 2;
int y = 3;
int s = x * y;
assertEquals(6, s);
}
```

assertEquals()方法中前一个值是预期的值，第二个值是参数的真实值。通过这个方法，可以判断运行结果是否正确。

(4) 进行测试。完成上述步骤以后，右击类名，选择 Run As|Android Juint Test 进行测试，测试结果如图 3-13 所示。如果运行正确，JUnit 窗口会显示绿色条；如果运行错误，JUnit 窗口会显示红色条。单击出错的方法，可定位到出错的源代码。

JUnit 不需要关注控制层，当业务层逻辑写好以后就可以进行测试了。这可以保证简单高效地开发应用程序。

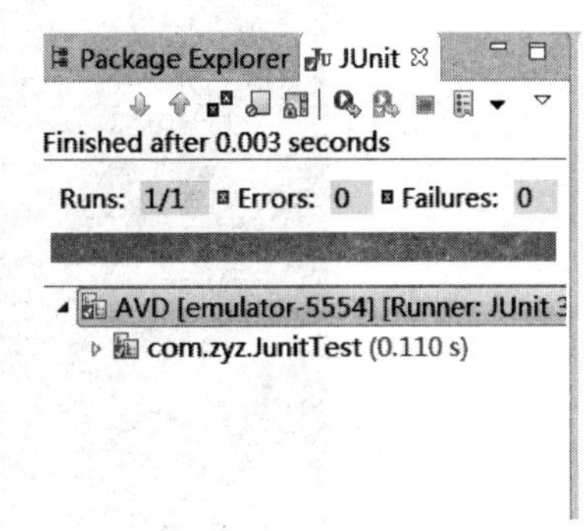

图 3-13　LogCat JUnit 运行结果

3.4　资源调用

在 Android 开发过程中，仅仅编写代码是远远不够的。要开发一个功能完备、实用的 Android 应用离不开各种外部资源的支持，例如字符串、图片、音频和视频等。这些外部资源也将作为应用的一部分，被编译到应用程序中。

在 Android 项目中，外部资源一般被放在 res 和 assets 文件夹中。

存放在 res 中的资源一般是可以通过 Android 的 R 类直接访问的资源，例如布局、图片和各种 xml 文件。

assets 文件夹里存放不能被 Android 程序直接访问的资源，例如视频文件。

Android 存放的各种资源如表 3-3 所示。

表 3-3 Android 资源目录

目录结构	存放的资源类型
res/anim	动画文件
res/drawable	图片文件
res/layout	布局文件
res/values	各种 xml 资源文件
strings. xml	字符串文件
arrays. xml	数组文件
colors. xml	颜色文件
dimens. xml	尺寸文件
styles. xml	样式文件
res/xml	任意的 xml 文件
res/raw	直接复制到设备中的原生文件
res/menu	菜单文件

将代码与资源分离能够大幅提高程序的可维护性。例如在下面这个国际化例子中通过配置不同的资源，可以在不同的语言环境下显示不同的内容而不需要修改代码。如图 3-14 和图 3-15 所示。

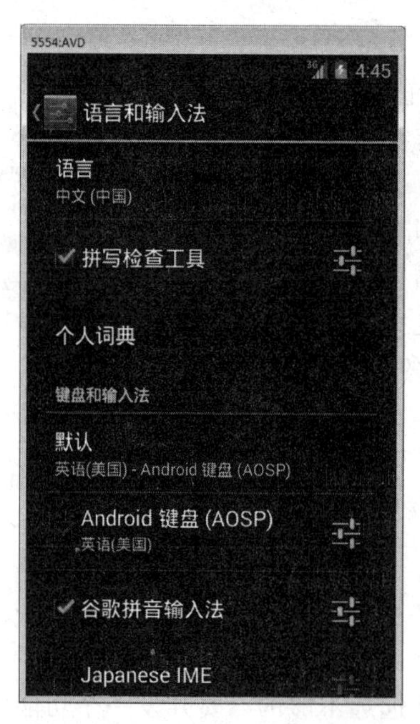

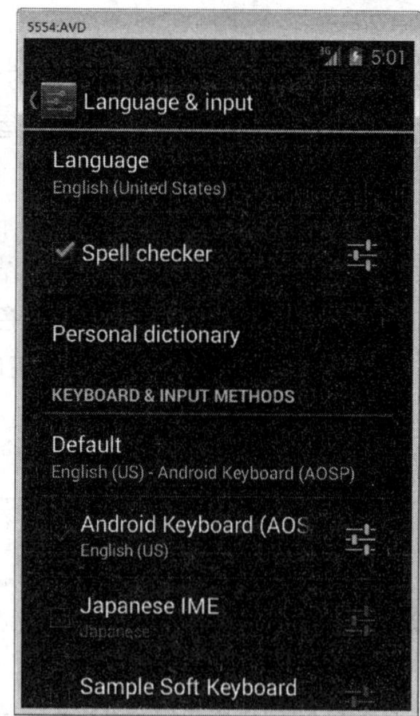

图 3-14 不同的语言环境

这两种效果使用的是同一个布局文件，其代码如下所示：

```
<RelativeLayout xmlns:android = "http://schemas.android.com/apk/res/android"
    xmlns:tools = "http://schemas.android.com/tools"
    android:layout_width = "match_parent"
    android:layout_height = "match_parent"
    tools:context = " ${relativePackage}. ${activityClass}" >
```

图 3-15 不同的效果显示

```
    <ImageView
        android:id="@+id/imageView1"
        android:layout_width="wrap_content"
        android:layout_height="wrap_content"
        android:src="@drawable/picture" />
     <TextView
        android:layout_width="wrap_content"
        android:layout_height="wrap_content"
        android:layout_below="@+id/imageView1"
        android:text="@string/hello_world" />
</RelativeLayout>
```

那么为什么会在不同的语言环境下显示出不同的内容呢？

这就是引入外部资源的好处了。

可以看到在< TextView >标签下有这样的代码：

```
android:text="@string/hello_world"
```

这说明 TextView 控件引用了 strings.xml 文件下的名称为 hello_world 的字符串的资源。

同理，看到< ImageView >标签下有这样的代码：

```
android:src="@drawable/picture"
```

这是图片控件引用 drawable 文件夹下的图片 picture.jpg。

前面说明了资源引用的过程，这里再介绍一下如何实现国际化。首先，在 AVD 中设置不同的语言，然后创建相应的资源文件夹。两个 drawable 文件夹分别命名为 drawable-zh-rCN 和 drawable-en-rUS。将两张不同的龙的图片都命名为 picture. jpg 并分别放在这两个文件夹下，如图 3-16 所示。

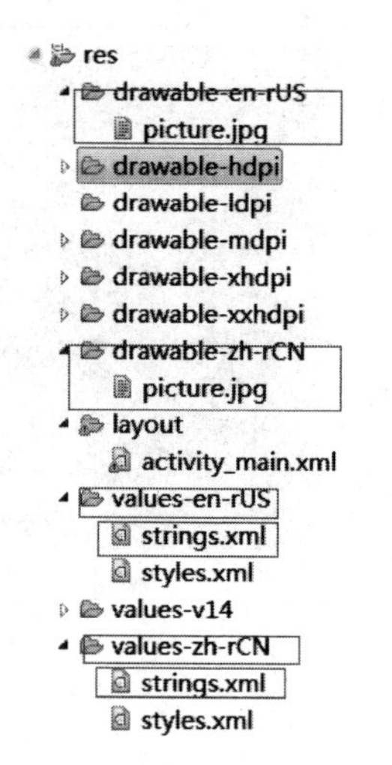

图 3-16 资源目录

同理将不同的 strings. xml 放在 values-zh-rCn、values-en-rUS 文件夹下。

values-zh-rCn 文件下的代码如下所示：

```
<?xml version = "1.0" encoding = "utf - 8"?>
< resources >
< string name = "app_name">国际化</string >
< string name = "hello_world">这是中国龙</string >

</resources >
```

只要改变< string name="hello_world">这是中国龙</string >，两个标签之间的文字，@string/hello_world 引用的文字就随之改变了。

3.5 本章小结

本章介绍了 Activity，讲解了 Activity 的生命周期以及如何使用 LogCat 观察生命周期的过程。读者应学会如何创建一个 Activity，如何使用 JUnit 进行单元测试，以及如何引用项目中的外部资源，了解如何在不同的语言环境下完成不同的显示效果。

3.6 练习题

一、填空题

1. ________、________、________方法在 Activity 初次创建时会被调用。
2. Android 的四大组件分别是________、________、________、________。
3. 图片文件在编写 Android 应用时，是通过________标签调用的。

二、选择题

1. 以下目录用来存放 strings. xml 的是(　　)。

   A. src 目录　　B. libs 目录
   C. assets 目录　　D. res/values 目录

2. Activity 的生命周期中标志 Activity 被销毁方法的是(　　)。

   A. onCreate()　　B. onDestroy()
   C. onStart()　　D. onPause()

三、简答题

1. 简单描述 Activity 的生命周期。
2. 如何创建一个 JUnit 单元测试。
3. 如何使 LogCat 只显示错误信息。

四、编程题

编写一个 Activity,显示一张图片。

第4章 常见的UI控件

本章重点

- 基本控件：TextView,Button,EditText,ImageView
- 弹出框：ProgressBar,AlertDialog,ProgressDialog,Toast
- ListView的基本使用
- 自定义控件的使用

前面学习了如何创建一个Activity,现在来学习如何给Activity添加内容,让Activity更加丰富。Android中有很多方式可以编写程序界面。我们可能用过Adobe Dreamweaver的可视化界面,它可以以拖动控件的方式来编写布局。其实,Eclipse中也有相应的可视化编辑器,和Adobe Dreamweaver的可视化界面用法差不多,都是可以直接拖动控件的,并且能在可视化界面修改控件属性。不过并不推荐使用这种方法,因为可视化编辑工具不能让你很好地了解界面背后的复杂原理。但是这个功能还是有它的作用的,那就是在代码编辑之后用来进行界面预览,毕竟Android每次调试安装都是很费事的。

下面介绍常见的UI控件。

4.1 基本控件的使用方法

在使用Android手机的时候,都看到过很多好看、功能强大的控件,这些功能实现起来其实很简单。Android提供了很多的控件,例如TextView,Button,EditText,ImageView这些常见的控件,还有很多扩展的控件、自定义的控件等等。下面来学习常见控件的使用。

4.1.1 TextView

TextView是最基本的一个控件,前面案例已经使用过它。它的主要功能是显示一段文本信息,可以是图文混排的,也可以是纯文本的。布局文件实例：

```
<?xml version = "1.0" encoding = "utf - 8"?>
< RelativeLayout xmlns:android = "http://schemas.android.com/apk/res/android"
    xmlns:tools = "http://schemas.android.com/tools"
    android:layout_width = "match_parent"
    android:layout_height = "match_parent"
    android:paddingBottom = "@dimen/activity_vertical_margin"
```

```
    android:paddingLeft = "@dimen/activity_horizontal_margin"
    android:paddingRight = "@dimen/activity_horizontal_margin"
    android:paddingTop = "@dimen/activity_vertical_margin">
    <TextView
        android:id = "@ + id/tv_test"
        android:layout_width = "wrap_content"
        android:layout_height = "wrap_content"
        android:background = " # e22017"
        android:text = "这是一个文本框"
        android:textColor = " # 87e266"
        android:textSize = "33sp" />
</RelativeLayout>
```

最外边的 RelativeLayout 表示相对布局，布局效果如图 4-1 所示。在 TextView 中定义的属性说明如下：

- android:id：设置控件的唯一标识。android:id="@+id/tv_test"会在 R 文件中创建一个新的唯一标识。
- android:layout_width：设置控件的宽度。android:layout_width ="wrap_content"设置该组件大小可根据内容来修改。

该属性还有其他两个参数分别是 fill_parent 和 match_parent，这两个参数意思是一样的，都是填充父类窗口。从 Android 2.2(API 8)开始可以直接用 match_parent 来代替 fill_parent，不过 Android 2.2 以后的版本 fill_parent 仍然可以使用，其含义都是一样的，都是填满整个父窗口。

- android:layout_height：设置控件的高度。属性值与设置 android:layout_width 一样。
- android:text：设置该控件要显示的文本信息。
- android:textSize：设置文本字体大小。
- android:textColor：设置文本颜色。
- android:background：设置背景颜色。

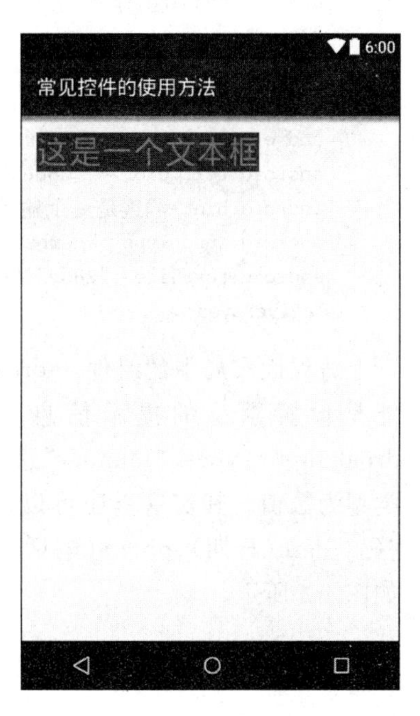

图 4-1 布局效果(一)

4.1.2 EditText

EditText 是文本编辑框，它的父类是文本框 TextView，所以它可以调用 TextView 大部分的 xml 属性和方法。EditText 与 TextView 最大的区别是：EdtiText 可以接收用户输入。

如果想设置 EditText 控件中的文本不让别人看到，那就需要用到 inputType 这个属性，可以将 EditText 设置为指定类型的输入组件。

inputType 常用的属性值如下：

- text：普通文本。
- textMultiLine：多行文本。

- textPassword：密码格式。
- Number：数字格式。

还有其他格式例如小数点、URI格式、短消息格式、长消息格式等。

布局文件实例如下：

```
<?xml version = "1.0" encoding = "utf - 8"?>
<RelativeLayout xmlns:android = "http://schemas.android.com/apk/res/android"
    xmlns:tools = "http://schemas.android.com/tools"
    android:layout_width = "match_parent"
    android:layout_height = "match_parent"
    android:paddingBottom = "@dimen/activity_vertical_margin"
    android:paddingLeft = "@dimen/activity_horizontal_margin"
    android:paddingRight = "@dimen/activity_horizontal_margin"
    android:paddingTop = "@dimen/activity_vertical_margin">
    <EditText
    android:id = "@ + id/edt_test"
    android:layout_width = "388dp"
    android:layout_height = "wrap_content"
    android:textColor = "#88e355"
    android:hint = "这是一个输入框"
    android:inputType = "number"
    android:textSize = "28sp" />
</RelativeLayout>
```

上边界面布局中的组件android:hint这个属性主要的功能是设置默认的提示信息："这是一个输入框"；android:inputType="number"这个属性主要是设置输入的类型为数值。其实这里还可以设置如numberPassword(密码)、date(日期)、phone(电话号码)等属性。该布局效果如图4-2所示。

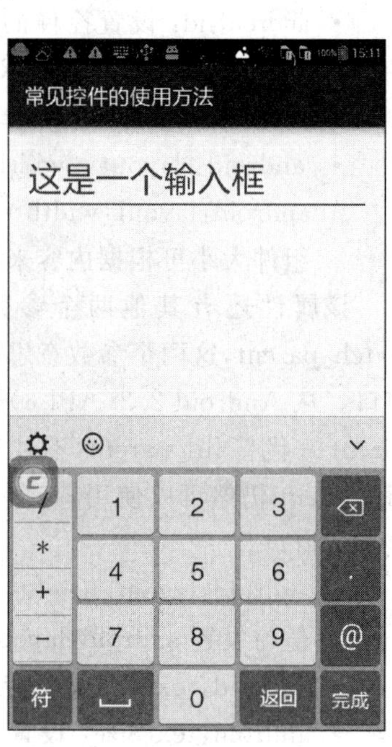

图4-2 布局效果(二)

4.1.3 Button

Button是一个按钮控件。跟EditText一样，其父类也是TextView，所以Button也能调用TextView大部分的xml属性和方法。

```
<?xml version = "1.0" encoding = "utf - 8"?>
<LinearLayout xmlns:android = "http://schemas.android.com/apk/res/android"
    xmlns:tools = "http://schemas.android.com/tools"
    android:layout_width = "match_parent"
    android:layout_height = "match_parent"
    android:orientation = "vertical" >

    <TextView
        android:id = "@ + id/tv_test"
        android:layout_width = "wrap_content"
        android:layout_height = "wrap_content"
        android:background = "#e22017"
```

```
        android:text = "这是一个文本框"
        android:textColor = "#87e266"
        android:textSize = "33sp" />

    <Button
        android:id = "@ + id/btn_test"
        android:layout_width = "match_parent"
        android:layout_height = "wrap_content"
        android:layout_margin = "16dp"
        android:gravity = "left"
        android:onClick = "onClickXml"
        android:padding = "12dp"
        android:text = "这是一个按钮" />

</LinearLayout>
```

布局文件中的 android:padding 属性与 HTML 中的 paddding 类似，主要指该控件内部内容(如文本)距离该控件边缘的距离；android:layout_margin 指该控件距离其父控件边缘的距离；android:gravity 主要是对控件里的元素来说的，用来控制元素在该控件里的显示位置。默认是左上对齐，这里设置的是左对齐。

加入 Button 之后的界面效果如图 4-3 所示。在 TestActivity 中设置 Button 的单击事件，单击之后的效果如图 4-4 所示。

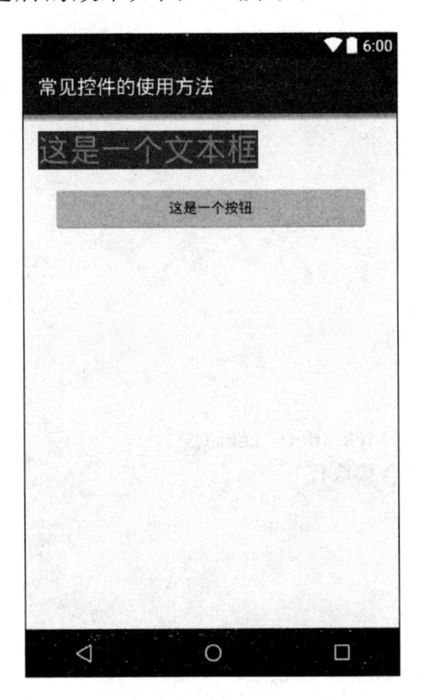

图 4-3 加入 Button 之后的界面效果

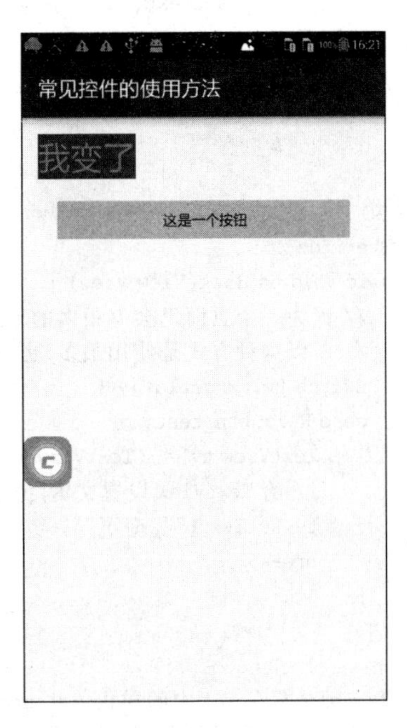

图 4-4 最终效果

然后可以在 TestActivity 中的 Button 的单击事件注册一个监听器，如下所示：

```
package com.qsd;

import android.app.Activity;
import android.os.Bundle;
```

```
import android.view.View;
import android.widget.Button;
import android.widget.TextView;

import com.qsd.ch4_1.R;

public class ButtonActivity extends Activity implements View.OnClickListener {
    private Button mButton;

    @Override
    protected void onCreate(Bundle savedInstanceState) {
        super.onCreate(savedInstanceState);
        setContentView(R.layout.activity_button);
        // 第一种方式
        // 获取 Button 按钮
        mButton = (Button) findViewById(R.id.btn_test);
        // 给 Button 设置监听事件
        mButton.setOnClickListener(new View.OnClickListener() {
            @Override
            public void onClick(View view) {
                TextView tv = (TextView) findViewById(R.id.tv_test);
                // 给 TextView 设置文本内容
                tv.setText("我变了");
            }
        });

    }

    // 第二种方式
    @Override
    public void onClick(View view) {
        // 因为一个页面里会有很多的单击事件,
        // 所以这种方式是使用最多,也是最常用的
        switch (view.getId()) {
        case R.id.btn_test:
            TextView tv = (TextView) findViewById(R.id.tv_test);
            // 给 TextView 设置文本,原来的文本会被取代
            tv.setText("我变了");
            break;
        }
    }

    //第三种是写在 xml 中的单击方法 android:onClick = "onClick",
    // 这种方法与第二种类似,但是没有@Override
    public void onClickXml(View view) {

        switch (view.getId()) {
        case R.id.btn_test:
            TextView tv = (TextView) findViewById(R.id.tv_test);
            // 给 TextView 设置文本,原来的文本会被取代
            tv.setText("我变了");
```

```
                break;
            }
        }
    }
```

4.1.4 ImageView

ImageView是一个展示任意图片的控件，它的父类是View类，所以ImageView适用任何布局，它的主要功能就是显示图片。

```
<?xml version = "1.0" encoding = "utf - 8"?>
< LinearLayout xmlns:android = "http://schemas.android.com/apk/res/android"
    xmlns:tools = "http://schemas.android.com/tools"
    android:layout_width = "match_parent"
    android:layout_height = "match_parent"
    android:orientation = "vertical" >

    < ImageView
        android:id = "@ + id/iv_test"
        android:layout_width = "wrap_content"
        android:layout_height = "wrap_content"
        android:src = "@drawable/ic_launcher" />

</LinearLayout >
```

这里使用android:src属性给ImageView设置了一张在drawable文件夹下的图片，并将ImageView的高和宽都设置成了wrap_content，保证控件能自动适应图片的大小。效果图如图4-5所示。

图4-5 设置效果

4.2 常见的弹出框基本使用

4.2.1 ProgressBar

ProgressBar是一个进度条控件,它的父类跟ImageView一样都是View,用于显示一个条,表示该操作的进展程度。当该操作有进展时,应用程序可以改变条的长度。在进度条上还有一个可显示的辅助进度,这对于显示中间的进度很有用,例如在流回放进度中缓冲情况。

进度条也可以不确定。在不确定模式下,进度条显示循环动画,不显示进度。当任务的长度未知时,程序可以使用此模式。不确定的进度条可以是旋转轮或水平条。代码如下:

```
<LinearLayout xmlns:android="http://schemas.android.com/apk/res/android"
    xmlns:tools="http://schemas.android.com/tools"
    android:layout_width="match_parent"
    android:layout_height="match_parent"
    android:orientation="vertical"
    tools:context=".TestActivity">

    <ProgressBar
        android:id="@+id/pgb_test"
        android:layout_width="wrap_content"
        android:layout_height="wrap_content"
        android:layout_gravity="center" />

    <ProgressBar
        android:id="@+id/pgb_test1"
        android:layout_width="match_parent"
        android:layout_height="wrap_content"
        android:layout_gravity="center"
        android:max="100"
        android:visibility="gone"
        style="?android:attr/progressBarStyleHorizontal" />
</LinearLayout>
```

代码中,android:layout_gravity属性是控制控件本身的对齐方式,用来控制本身相对于父控件中的位置;android:visibility是所有的Android控件都有的属性,它只有三种可选的属性值,分别是visible、invisible和gone,主要用来设置控件的显示和隐藏。

(1) 可见(visible)。

```
XML文件: android:visibility="visible"
Java代码: view.setVisibility(View.VISIBLE);
```

(2) 不可见(invisible)。

```
XML文件: android:visibility="invisible"
Java代码: view.setVisibility(View.INVISIBLE);
```

(3) 隐藏(gone)。

```
XML 文件: android:visibility = "gone"
Java 代码: view.setVisibility(View.GONE);
```

属性值 gone 和 invisible 实现的功能可以说是一样的，都是把控件隐藏让用户看不见，但是设置属性值为 invisible 时，等于在界面中设置了一个占位符，不管用不用都必须留着，而 gone 则不占用空间。

在图 4-6 中，设置的是圆形进度条。可以通过 sytle 属性将它设置成水平进度条，如改成 style="? android:attr/progressBarStyleHorizontal"。

XML 代码如下：

```
<?xml version = "1.0" encoding = "utf - 8"?>
<LinearLayout xmlns:android = "http://schemas.android.com/apk/res/android"
    xmlns:tools = "http://schemas.android.com/tools"
    android:layout_width = "match_parent"
    android:layout_height = "match_parent"
    android:orientation = "vertical" >

    <TextView
        android:layout_width = "match_parent"
        android:layout_height = "wrap_content"
        android:gravity = "left" />

    <ProgressBar
        android:id = "@ + id/pgb_test"
        android:layout_width = "wrap_content"
        android:layout_height = "wrap_content"
        android:layout_gravity = "center" />

    <ProgressBar
        android:id = "@ + id/pgb_test1"
        style = "?android:attr/progressBarStyleHorizontal"
        android:layout_width = "match_parent"
        android:layout_height = "wrap_content"
        android:layout_gravity = "center"
        android:max = "100"
        android:visibility = "visible" />

    <Button
        android:id = "@ + id/btn_test"
        android:layout_width = "200dp"
        android:layout_height = "100dp"
        android:layout_gravity = "center_horizontal"
        android:layout_marginTop = "20dp"
        android:text = "增加进度条" />

</LinearLayout>
```

后台代码如下：

```
package com.qsd;

import android.app.Activity;
import android.os.Bundle;
import android.view.View;
import android.widget.ProgressBar;

import com.qsd.ch4_2.R;

public class ProgressBarActivity extends Activity {
    private ProgressBar mProgress;

        @Override
        protected void onCreate(Bundle savedInstanceState) {
            super.onCreate(savedInstanceState);
            setContentView(R.layout.activity_progress_bar);
            //获取控件
            mProgress = (ProgressBar) findViewById(R.id.pgb_test1);
            //获取按钮并设置单击事件
            findViewById(R.id.btn_test).setOnClickListener(new View.OnClickListener() {
                @Override
                public void onClick(View view) {
                    // 取得运行进度值
                    int progress = mProgress.getProgress();
                    progress = progress + 10;
                    //把算好的进度值设置回去
                    mProgress.setProgress(progress);
                    if (progress == 100) {//如果进度条达到 100 的时候
                        //进度条设置成隐藏
                        mProgress.setVisibility(View.GONE);
                    }
                }
            });

        }
}
```

每单击一次按钮，可获取进度条的当前进度，然后在现有的进度上加上 10 作为更新后的进度（见图 4-6、图 4-7）。ProgressBar 还有几种其他样式，可以自己去尝试一下。

4.2.2 AlertDialog

AlertDialog 是一个特别的对话框，在对话框里可以显示一个、两个或三个按钮。它的父类是 Dialog。一般使用不会实例化 Dialog，而是实例化 Dialog 的子类。我们知道 Android 系统界面布局与 Photoshop 中的图层一样，是一层层地分别写上去的，是有先后顺序的，但是当对话框显示的时候它是置于所有控件之上的。一般 AlertDialog 都是用于提示

图 4-6　圆形进度条

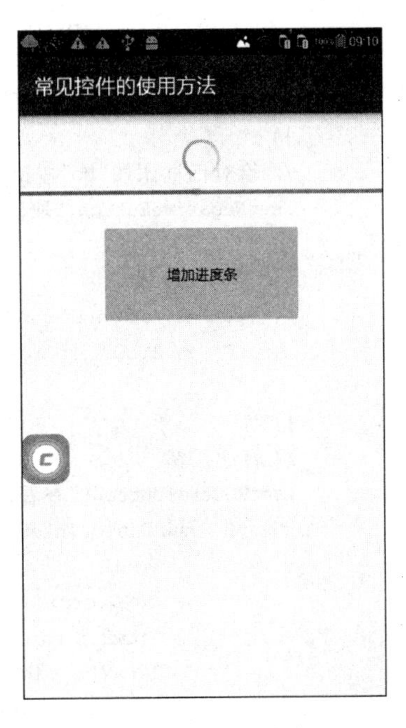

图 4-7　水平进度条

一些信息，例如退出提示、删除操作提示等。下面来学习一下它的简单使用，代码如下：

```
package com.qsd;

import android.app.Activity;
import android.app.AlertDialog;
import android.app.Dialog;
import android.content.DialogInterface;
import android.os.Bundle;

import com.qsd.ch4_2.R;

public class AlertDialogActivity extends Activity {

@Override
protected void onCreate(Bundle savedInstanceState) {
    super.onCreate(savedInstanceState);
    Dialog alertDialog = new AlertDialog.Builder(this)
    // 为对话框设置标题
            .setTitle("确定删除?")
            // 为对话框设置内容
            .setMessage("您确定删除该条信息吗?")
            // 为对话框设置图标
            .setIcon(R.drawable.ic_launcher)
            // 给对话框添加"Yes"按钮
            .setPositiveButton("确定", new DialogInterface.OnClickListener() {

                @Override
```

```
                public void onClick(DialogInterface dialog, int which) {
                    // TODO Auto-generated method stub
                }
            })
            // 给对话框添加"No"按钮
            .setNegativeButton("取消", new DialogInterface.OnClickListener() {

                @Override
                public void onClick(DialogInterface dialog, int which) {
                    // TODO Auto-generated method stub
                }
            })
            // 普通按钮
            .setNeutralButton("查看详情",
                    new DialogInterface.OnClickListener() {

                        @Override
                        public void onClick(DialogInterface dialog,
                                int which) {
                            // TODO Auto-generated method stub
                        }
                    }).create();        // 创建对话框
        alertDialog.show();                     // 显示对话框
    }
}
```

首先，通过 AlertDialog. Builder 创建一个 AlertDialog 的实例，然后可以为这个对话框设置内容、标题及图标等属性，接下来调用不同的方法，最后显示出来。效果如图 4-8 所示。

图 4-8 显示的对话框

4.2.3 ProgressDialog

ProgressDialog 和 AlertDialog 有点相似，因为它的父类是 AlertDialog，所以说它们都可以在界面上弹出一个对话框，都能够屏蔽掉其他控件的交互能力。不同的是，ProgressDialog 会在对话框中显示一个进度条，用来提示用户当前操作比较耗时，让用户耐心等待。它的用法和 AlertDialog 也比较相似。代码如下所示：

```
import android.app.ProgressDialog;
import android.os.Bundle;
import android.support.v7.app.AppCompatActivity;

public class DialogTestActivity extends AppCompatActivity {

    @Override
    protected void onCreate(Bundle savedInstanceState) {
        super.onCreate(savedInstanceState);
        setContentView(R.layout.activity_dialog_test);
        ProgressDialog pd = new ProgressDialog(this);
        pd.setTitle("标题");                            //设置标题
        pd.setMessage("加载中,请稍候……");               //设置提示信息
        pd.setCancelable(true);                         // 设置是否可以通过单击 Back 键取消
        pd.setCanceledOnTouchOutside(false);            //设置在单击 Dialog 外是否取消 Dialog 进度条
        pd.show();                                      //显示 ProgressDialog
    }
}
```

这里面要注意 setCancelable 这个属性，当设置参数是 false 的时候，一定要调用 dismiss() 方法来关闭。

对话框效果如图 4-9 所示。

图 4-9 显示的效果

4.2.4 Toast

Toast 是一个包含用户快速信息的视图。和 Dialog 不一样的是，Toast 永远不会获得焦点，无法被单击。Toast 类的作用就是既能向用户传递信息，又不会成为焦点。而且 Toast 显示的时间有限，Toast 会在用户设置的显示时间后自动消失。

activity_toast.xml 代码如下：

```
<?xml version="1.0" encoding="utf-8"?>
<LinearLayout xmlns:android="http://schemas.android.com/apk/res/android"
    xmlns:tools="http://schemas.android.com/tools"
    android:layout_width="match_parent"
    android:layout_height="match_parent"
    android:orientation="vertical"
    tools:context=".TestActivity" >

    <Button
        android:id="@+id/btn_toast"
        android:layout_width="200dp"
        android:layout_height="100dp"
        android:layout_gravity="center_horizontal"
        android:layout_marginTop="20dp"
        android:text="单击弹出提示信息" />

</LinearLayout>
```

XML 代码很简单，就只放了一个按钮，下面是 ToastActivity.java 代码。

```
package com.qsd;

import android.app.Activity;
import android.os.Bundle;
import android.view.View;
import android.view.View.OnClickListener;
import android.widget.Button;
import android.widget.Toast;

import com.qsd.ch4_2.R;

public class ToastActivity extends Activity {

@Override
protected void onCreate(Bundle savedInstanceState) {
    super.onCreate(savedInstanceState);
    setContentView(R.layout.activity_toast);
    // 获取 Button 对象
    Button button = (Button) findViewById(R.id.btn_toast);
    // 给 Button 对象设置单击事件
    button.setOnClickListener(new OnClickListener() {

        @Override
        public void onClick(View v) {
            Toast.makeText(ToastActivity.this, "这是一个 Toast",
```

```
                        Toast.LENGTH_LONG).show();

            }
        });

    }
}
```

代码中 Toast.makeText()方法里面的第一个参数是上下文环境，第二个参数是要显示的信息，第三个参数是信息显示的时长。时长参数有两个，一个是 Toast.LENGTH_LONG（或者 1）表示较长的时间显示，一个是 Toast.LENGTH_SHORT（或者 0）表示较短的时间显示。显示效果如图 4-10 所示。

图 4-10 显示效果

4.3 ListView 的基本使用

ListView 是 Android 软件开发中使用频率最高也是最难使用的控件，主要用于垂直显示列表项。单独写一个 ListView 一般需要写一个列表项 xml、要加载的数据、一个 Adapter 以及相对应的单击方法。

4.3.1 ListView 简单使用

首先在 layout 下新建一个 activity_list_view.xml。在布局中加入 ListView 控件，设一个 id 为 lv_test，宽度和高度选择填充父窗体，给 ListView 加一个分割线，分割线的颜色是 #E60000，分割线的粗细设置为 2dp。代码如下：

```
<?xml version = "1.0" encoding = "utf - 8"?>
< RelativeLayout xmlns:android = "http://schemas.android.com/apk/res/android"
xmlns:tools = "http://schemas.android.com/tools"
android:layout_width = "match_parent"
android:layout_height = "match_parent">
```

```
    <ListView
        android:id="@+id/lv_test"
        android:layout_width="match_parent"
        android:layout_height="match_parent"
        android:divider="#E60000"
        android:dividerHeight="2dp"></ListView>
</RelativeLayout>
```

接下来,编写 ListViewActivity 的代码,如下所示:

```
public class ListViewActivity extends Activity {
private String[] data = {"1", "2", "3", "4", "5", "6", "7", "8", "9", "10",
"11", "12"};
@Override
protected void onCreate(Bundle savedInstanceState) {
    super.onCreate(savedInstanceState);
    setContentView(R.layout.activity_list_view);
    //创建 ArrayAdapter 对象
    ArrayAdapter<String> mAdapter = new ArrayAdapter<String>(this,
            android.R.layout.simple_expandable_list_item_1, data);
    ListView mlv = (ListView) findViewById(R.id.lv_test);
    //为 ListView 设置 Adapter
    mlv.setAdapter(mAdapter);
}
}
```

上边程序中的 ArrayAdapter 是一个由任意对象数组支持的具体的适配器。默认情况下,这个类所需的资源 ID 会引用一个 TextView。

ArrayAdapter 中有三个参数,这三个参数分别是:

- Context:当前上下文,它表示应用程序环境的全局信息的接口。这是一个抽象类,其实现由 Android 系统提供。它允许访问特定于应用程序的资源和类,以及对应用程序级操作(如启动活动、广播和接收意图等)的调用。
- textViewResourceId:包含要在实例化视图时使用的布局的布局文件的资源 ID,我们引用了系统自带的一个 TextView。
- Data:要填充的数据。

android.R.layout.simple_expandable_list_item_1 是 Android 内置的布局文件,里面只有一个 TextView,用于简单地显示一段文字。上下文和数据参数已放到 ArrayAdapter 中了,之后再调用 ListView 中的 setAdapter 方法,关联新创建的 ArrayAdapter 对象,这样 ListView 和数据之间的关联就建立完成了。效果如图 4-11 所示,可以通过拖动来查看数据。

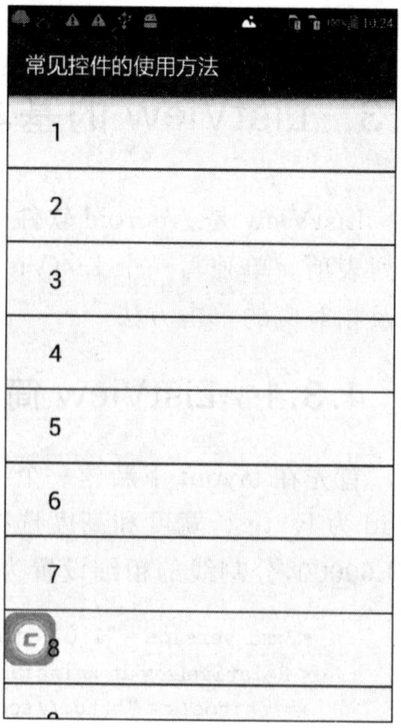

图 4-11 显示效果

4.3.2 ListView 使用进阶

现在使用的 ListView 功能非常丰富，但如果只是用来显示文本，则应自定义图片加文字的 ListView。

第一步，新建 item_number. xml，主要是替换 ListView 要填充的资源，在 layout 目录下的 item_number. xml 代码如下：

```
<?xml version = "1.0" encoding = "utf - 8"?>
< LinearLayout xmlns:android = "http://schemas.android.com/apk/res/android"
android:layout_width = "match_parent"
android:layout_height = "match_parent"
android:orientation = "horizontal">

< ImageView
    android:id = "@ + id/iv_num"
    android:layout_width = "wrap_content"
    android:layout_height = "wrap_content" />

< TextView
    android:id = "@ + id/tv_num"
    android:layout_width = "wrap_content"
    android:textSize = "20sp"
    android:layout_gravity = "center_vertical"
    android:textColor = " #20ab1b"
    android:layout_height = "wrap_content" />
</LinearLayout >
```

在这个布局中定义了一个 LinearLayout，方向水平；又定义了一个 ImageView，用于显示图片；定义了一个 TextView 用于显示文字。

第二步，创建一个 NumberInfo 的实体类，作为 Listview 适配器的适配类型。代码如下所示：

```
public class NumberInfo {
private int id;
private String num;
private int imageId;
public int getId() {
    return id;
}
public void setId(int id) {
    this.id = id;
}
public String getNum() {
    return num;
}
public void setNum(String num) {
    this.num = num;
}
```

```
    public int getImageId() {
        return imageId;
    }
    public void setImageId(int imageId) {
        this.imageId = imageId;
    }
}
```

NumberInfo类中有三个字段,id是主键,num表示要显示的数字,imageId表示相对应图片的资源的id。

第三步,再创建一个自定义的NumberAdapter,这个适配器继承自BaseAdapter。代码如下:

```
    public class NumberAdapter extends BaseAdapter {
    Context mcontext;
    List<NumberInfo> minfos;

    public NumberAdapter(Context context, List<NumberInfo> infos) {
        this.mcontext = context;
        this.minfos = infos;
    }

    @Override
    public int getCount() {
        //获取此适配器中数据集中的条目数
        return minfos.size();
    }

    @Override
    public Object getItem(int i) {
        //获取数据集中与指定索引对应的数据项
        return minfos.get(i);
    }

    @Override
    public long getItemId(int i) {
        //取在列表中与指定索引对应的行 id
        return minfos.get(i).getId();
    }

    @Override
    public View getView(int position, View convertView, ViewGroup parent) {
        // 使用自定义的 list_items 作为 Layout
        convertView = LayoutInflater.from(mcontext).inflate(R.layout.item_number, null);
        // 初始化布局中的元素
        ImageView mImageView = (ImageView) convertView.findViewById(R.id.iv_num);
        TextView mTextView = (TextView) convertView.findViewById(R.id.tv_num);
    //设置要显示的图片
        mImageView.setImageResource(minfos.get(position).getImageId());
    //设置要显示的文字
```

```
        mTextView.setText(minfos.get(position).getNum());
    //返回布局
        return convertView;
    }
}
```

NumberAdapter重写了父类的一组构造方法，将上下文和数据都传递进来，另外又重写了四个方法：getCount()，getItem()，getItemId()和getView()。其中的getView()方法是获取一个视图，显示数据集中指定位置的数据。在方法中手动创建的视图，用的是第一步创建的XML。当每次显示子项时getView()方法都会被调用一次。getCount()也非常重要，如果getCount返回0的话，是不显示数据的。

最后一步，把ListViewActivity中的代码修改一下，如下所示：

```
public class ListViewActivity extends Activity {
List<NumberInfo> minfos = new ArrayList<NumberInfo>();

@Override
protected void onCreate(Bundle savedInstanceState) {
    super.onCreate(savedInstanceState);
    setContentView(R.layout.activity_list_view);
    initData();                              //初始化数据
    NumberAdapter mAdapter = new NumberAdapter(this, minfos);
    ListView mlv = (ListView) findViewById(R.id.lv_test);
    mlv.setAdapter(mAdapter);
}

private void initData() {
    for (int i = 0; i < 30; i++) {
        NumberInfo info = new NumberInfo();
        info.setId(i);
        info.setImageId(R.drawable.ic_launcher);
        info.setNum("这是数字" + i);
        minfos.add(info);
    }
}
}
```

可以看到，代码中添加了一个用于初始化所有的数据initData()方法。再看一下效果，如图4-12所示。

图4-12 显示效果

4.3.3 ListView使用优化

前面介绍了ListView的基本使用方式。有时如果数据多的话，会出现崩溃或运行效率很低的情况。如何避免呢？这就需要进行优化。下面先看代码：

```
    @Override
public View getView(int position, View convertView, ViewGroup parent) {
    View view;
```

```
        if (convertView == null) {
            // 使用自定义的 list_items 作为 Layout
            view = LayoutInflater.from(mcontext).inflate(R.layout.item_number, null);
        } else {
            view = convertView;
        }
        // 初始化布局中的元素
        ImageView mImageView = (ImageView) view.findViewById(R.id.iv_num);
        TextView mTextView = (TextView) view.findViewById(R.id.tv_num);
        mImageView.setImageResource(minfos.get(position).getImageId());
        mTextView.setText(minfos.get(position).getNum());
        return view;
    }
```

getView()中提供了一个 convertView,因为如果不使用缓存 convertView 的话,调用 getView 时每次都会重新创建 View,这样之前的 View 可能还没有销毁,加之不断新建 View 势必会造成内存泄露。

重用缓存 convertView 传递给 getView()方法可避免填充不必要的视图。Google 官方提示还可以通过使用 ViewHolder 模式来避免没有必要地调用 findViewById(),因为太多的 findViewById 也会影响性能。ViewHolder 是一个静态类,使用 ViewHolder 模式的好处是缓存了显示数据的视图(View),加快了 UI 的响应速度。在使用之前要先判断 convertView 是否为空。如果 convertView 为空,根据设计好的 List 的 Item 布局(XML)为 convertView 赋值,并生成一个 viewHolder 来绑定 converView 里面的由 XML 布局设置的各个 View 控件,再用 convertView 的 setTag 将 viewHolder 设置到 Tag 中,以便系统第二次引用 ListView 时从 Tag 中取出;如果 convertView 不为空,就会直接用 convertView 的 getTag()来获得一个 ViewHolder。具体实现代码如下。

```
@Override
public View getView(int position, View convertView, ViewGroup parent) {
    View view;
    ViewHolder viewHolder;
    if (convertView == null) {
        // 使用自定义的 list_items 作为 Layout
        view = LayoutInflater.from(mcontext).inflate(
                R.layout.item_number, null);
        viewHolder = new ViewHolder();
        // 初始化布局中的元素
        viewHolder.mImageView = (ImageView) view.findViewById(R.id.iv_num);
        viewHolder.mTextView = (TextView) view.findViewById(R.id.tv_num);
        view.setTag(viewHolder);
    } else {
        view = convertView;
        viewHolder = (ViewHolder) view.getTag();
    }

    viewHolder.mImageView.setImageResource(minfos.get(position).getImageId());
    viewHolder.mTextView.setText(minfos.get(position).getNum());
    return view;
}
```

```
/***
 * 用户对控件的实例进行缓存
 */
class ViewHolder {
    ImageView mImageView;
    TextView mTextView;
}
```

4.3.4 ListView 单击方法

介绍完了 ListView 的使用方法,接下来介绍一下它的单击方法。如果 ListView 的子项不能单击的话,这个控件就失去了它的存在价值了。

先看如下的一段代码。这段代码使用了一个 setOnItemClickListener()方法为 ListView 注册了一个监听器,当用户单击 ListView 中的任何一个子项时就会调用 onItemClick()方法,在这个方法中可以通过 position 参数判断用户单击的是哪一个子项,然后获取相对应的数据,并通过 Toast 将数据显示出来。效果如图 4-13 所示。

```
public class ListViewActivity extends Activity {
List<NumberInfo> minfos = new ArrayList<NumberInfo>();

@Override
protected void onCreate(Bundle savedInstanceState) {
    super.onCreate(savedInstanceState);
    setContentView(R.layout.activity_list_view);
    initData();                                  //初始化数据
    NumberAdapter mAdapter = new NumberAdapter(this, minfos);
    ListView mlv = (ListView) findViewById(R.id.lv_test);
    mlv.setAdapter(mAdapter);
    mlv.setOnItemClickListener(new AdapterView.OnItemClickListener() {
        @Override
        public void onItemClick(AdapterView<?> adapterView, View view, int i, long l) {
            Toast.makeText(ListViewActivity.this,
                    minfos.get(i).getNum(), Toast.LENGTH_LONG).show();
        }
    });
}

private void initData() {
    for (int i = 0; i < 30; i++) {
        NumberInfo info = new NumberInfo();
        info.setId(i);
        info.setImageId(R.drawable.ic_launcher);
        info.setNum("这是数字" + i);
        minfos.add(info);
    }
}
}
```

其实还有两种给 ListView 子项添加单击方法的做法,一种是在 xml 注册监听事件,另一种方法是让 Activity 类实现接口 implements AdapterView.OnItemClickListener,效果是一样的,在此不再赘述。

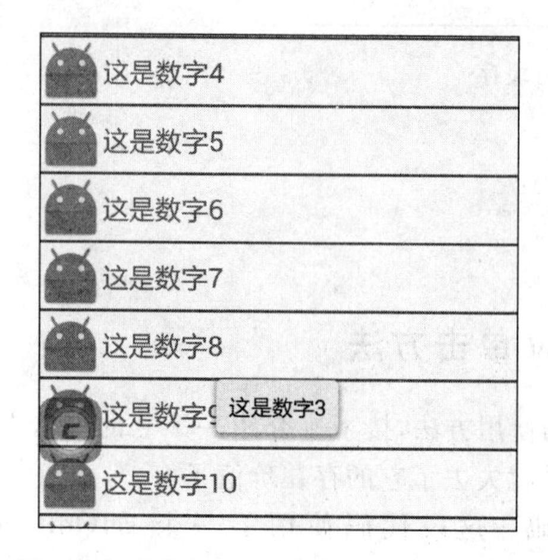

图 4-13 当单击“这是数字 3”时的显示效果

4.4 自定义控件

前面介绍了常见的 UI 控件，只是有的系统控件外观无法满足设计要求，因此，需要自定义开发一些控件，而且有很多界面是可以复用的。在学习自定义控件之前先学习一下它们的继承关系，如图 4-14 所示。

View 代表了用户界面的一块可绘制的区域。每个 View 在屏幕上占据一个矩形区域，负责绘图和事件处理。Android 中所有控件都继承自 android. view. View，是小部件的基类，用于创建交互式 UI 组件（按钮、文本字段等），其中 android. view. ViewGroup 是 View 的一个重要子类，绝大部分的布局都继承自 ViewGroup。

4.4.1 引用布局

大家在使用手机的时候都应该看到过每个界面都是有标题栏的，标题栏一般情况下都会带有一个或者两个按钮，用于响应用户返回或者其他操作。

新建一个 view_title. xml 布局文件，代码如下：

```
<?xml version = "1.0" encoding = "utf - 8"?>
<LinearLayout xmlns:android = "http://schemas.android.com/apk/res/android"
    android:layout_width = "match_parent"
    android:orientation = "horizontal"
    android:layout_height = "match_parent">

    <Button
        android:id = "@ + id/btn_back"
        android:layout_width = "wrap_content"
        android:layout_height = "wrap_content"
        android:text = "返回" />

    <TextView
        android:layout_width = "wrap_content"
```

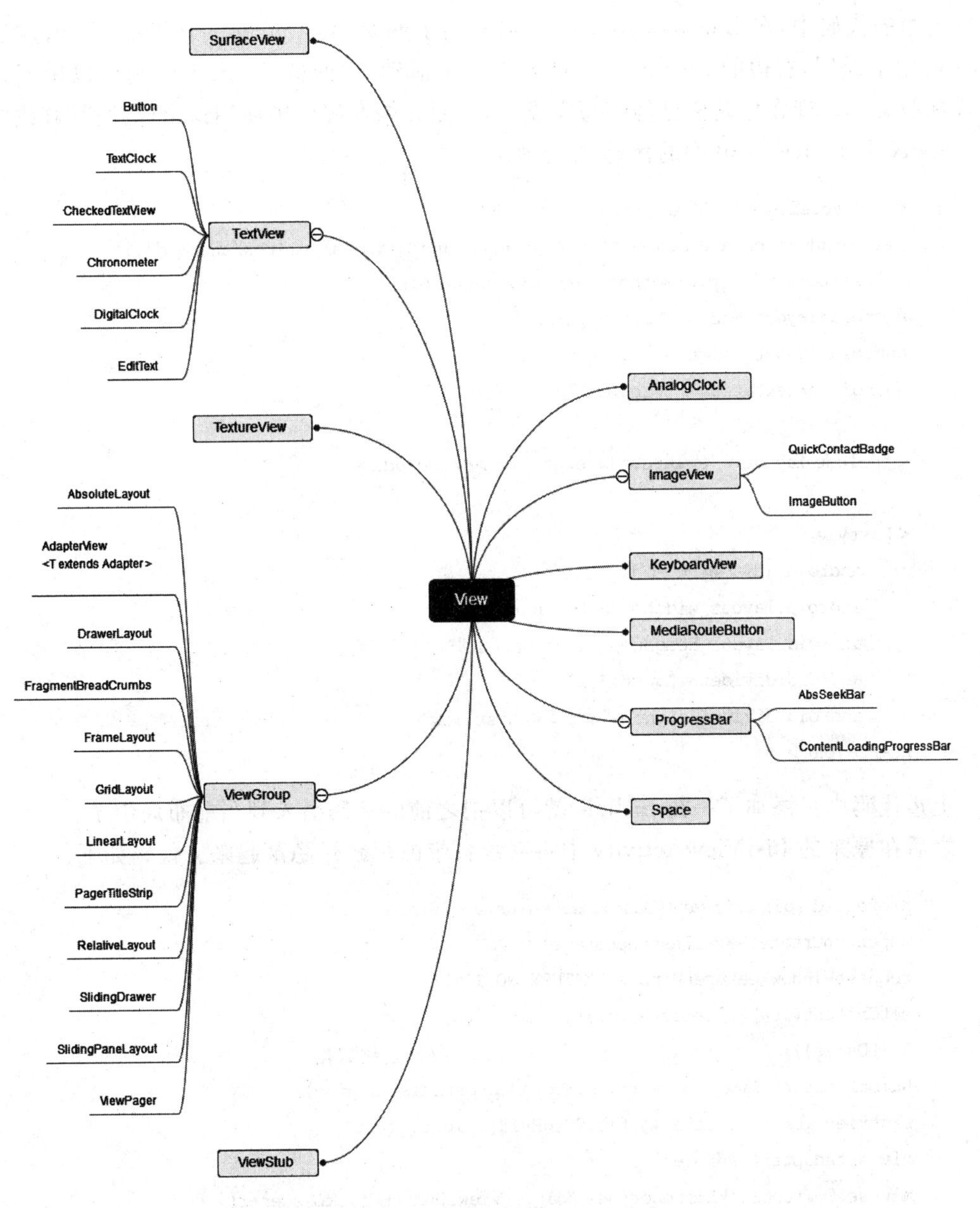

图 4-14 Android 中控件继承关系

```
        android:layout_height = "wrap_content"
        android:layout_weight = "1"
        android:gravity = "center"
        android:text = "这个是标题" />

    <Button
        android:id = "@ + id/btn_next"
        android:layout_width = "wrap_content"
        android:layout_height = "wrap_content"
        android:text = "下一个" />
</LinearLayout>
```

上边的代码中，在 LinearLayout 中分别添加了两个 Button 和一个 TextView，左边的 Button 用于返回，右边的 Button 可于进入下一个活动，中间的 TextView 则可以用于显示一段标题文本。现在标题栏已经编写完成了，下面开始在程序中使用这个标题栏，修改之前的 activity_list_view.xml 中的代码，如下所示：

```
<?xml version="1.0" encoding="utf-8"?>
<LinearLayout xmlns:android="http://schemas.android.com/apk/res/android"
    xmlns:tools="http://schemas.android.com/tools"
    android:layout_width="match_parent"
    android:layout_height="match_parent"
    android:orientation="vertical">

    <include layout="@layout/view_title"></include>

    <ListView
        android:id="@+id/lv_test"
        android:layout_width="match_parent"
        android:layout_height="match_parent"
        android:divider="#dd430b"
        android:dividerHeight="2dp"></ListView>
</LinearLayout>
```

上边代码中只添加了一行 include 就可以把之前的代码引入到当前布局中了。
之后在原来的 ListViewActivity 中将系统自带的标题栏隐藏起来。代码如下：

```
protected void onCreate(Bundle savedInstanceState) {
super.onCreate(savedInstanceState);
requestWindowFeature(Window.FEATURE_NO_TITLE);
setContentView(R.layout.activity_list_view);
initData();                                   //初始化数据
NumberAdapter mAdapter = new NumberAdapter(this, minfos);
ListView mlv = (ListView) findViewById(R.id.lv_test);
mlv.setAdapter(mAdapter);
mlv.setOnItemClickListener(new AdapterView.OnItemClickListener() {
    @Override
    public void onItemClick(AdapterView<?> adapterView, View view, int i, long l) {
        Toast.makeText(ListViewActivity.this,
                minfos.get(i).getNum(), Toast.LENGTH_LONG).show();
    }
});
}
```

注意，应该在 setContentView()之前添加 requestWindowFeature(Window.FEATURE_NO_TITLE)，因为如果不在 setContentView()之前添加的话就没什么作用了。效果如图 4-15 所示。

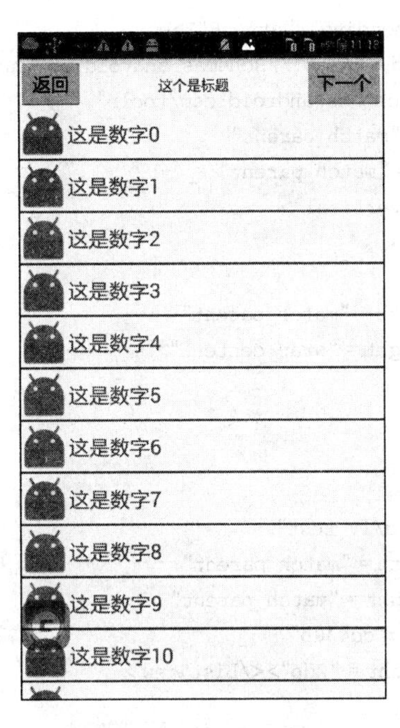

图 4-15 引用布局效果

4.4.2 创建自定义布局

如果想给引用布局添加单击事件，通常只能在引用 Activity 中添加。例如，标题栏中的“返回”按钮，不管在哪个 Activity 中这个按钮的功能都是“返回”。如何在一个布局中就解决问题呢？那就是自定义控件的解决方式。代码如下：

```
public class TitleLayout extends LinearLayout {
    public TitleLayout(Context context, AttributeSet attrs) {
        super(context, attrs);
LayoutInflater layoutInflater = LayoutInflater.from(context);
        layoutInflater .inflate(R.layout.view_title, this);
    }
}
```

新建一个 TitleLayou.java，它继承自 LinearLayout。对 LayoutInflater 已经见识过了，在前边就用它来完成 ListView 的子项布局文件。在 Activity 加载布局的时，通常是用 setContentView()方法来完成的。

通过 LayoutInflater 静态 from()方法给定的上下文获取 LayoutInflater 的对象。LayoutInflater 中有一个 inflate()方法，用到的是 inflate(int resource, ViewGroup root)方法。该方法第一个参数就是要加载的 resource，我们传入了布局 id(R.layout.view_title)。第二个参数是指给该布局的外部再嵌套一层父布局 root，如果不需要就直接传 null，这里指定为 this，也就是当前的 TitleLayout。这样就创建成功一个布局实例，在相对应的 xml 文件中将其添加到指定位置就行。

修改 activity_list_view.xml 中的代码，代码如下：

```
    <?xml version = "1.0" encoding = "utf - 8"?>
<LinearLayout xmlns:android = "http://schemas.android.com/apk/res/android"
    xmlns:tools = "http://schemas.android.com/tools"
    android:layout_width = "match_parent"
    android:layout_height = "match_parent"
    android:orientation = "vertical">

    <com.qsd.TitleLayout
        android:layout_width = "match_parent"
        android:layout_height = "wrap_content">

    </com.qsd.TitleLayout>

    <ListView
        android:id = "@ + id/lv_test"
        android:layout_width = "match_parent"
        android:layout_height = "match_parent"
        android:divider = "#dd430b"
        android:dividerHeight = "2dp"></ListView>
</LinearLayout>
```

添加自定义控件和使用普通控件的方式基本一样，只不过在添加自定义控件的时候需要指明控件的完整类名(例如 com.qsd.TitleLayout)，包名不能像清单文件那样可以省略，在 xml 中是不可以省略的。

然后在 TitleLayout 中为标题栏注册单击事件，代码如下：

```
public class TitleLayout extends LinearLayout {
public TitleLayout(Context context, AttributeSet attrs) {
    super(context, attrs);
    LayoutInflater.from(context).inflate(R.layout.view_title, this);
    //初始化控件
    initView();
}

private void initView() {
    Button mBack = (Button) findViewById(R.id.btn_back);
    Button mNext = (Button) findViewById(R.id.btn_next);
    final TextView mtitle = (TextView) findViewById(R.id.tv_title);
    mBack.setOnClickListener(new OnClickListener() {
        @Override
        public void onClick(View view) {
            mtitle.setText("您单击了返回按钮");
        }
    });
    mNext.setOnClickListener(new OnClickListener() {
        @Override
        public void onClick(View view) {
            mtitle.setText("您单击了下一个按钮");
        }
    });
}
```

这里写了一个 initView()方法，主要目的是使之显得更有条理。其中，分别获取了控件的实例并给两个按钮注册了单击事件，当单击按钮的时候在中间的 TextView 控件显示相应的提示信息。重新运行代码，单击“返回”按钮，效果如图 4-16 所示。

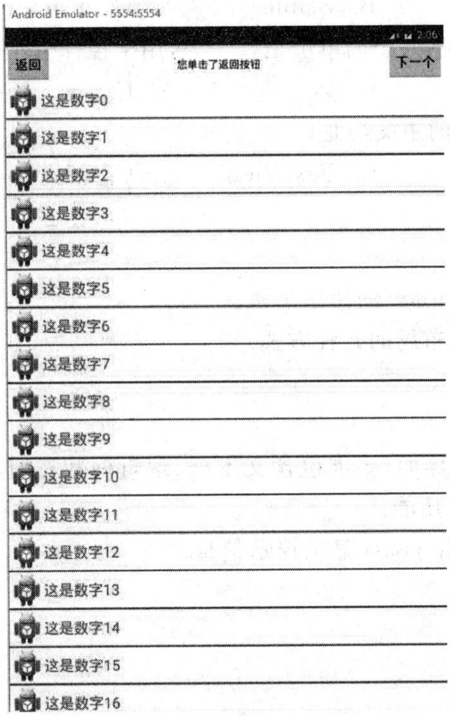

图 4-16　单击“返回”按钮后的效果

4.5　本章小结

虽然本章的内容很多，但是学习起来还是有很多乐趣的。本章依次介绍了基本布局的用法、ListView 的详细使用和自定义控件的使用，基本已经将重要的 UI 知识点覆盖了。不难发现，即使不使用可视化的编辑工具，而是全部使用 XML 的方式来编写界面，也不是太难的事情。本章的案例，只是介绍了每个控件的常用属性，其实还有很多属性可以设置，所以应该学会使用搜索工具和帮助文档扩展自己的知识面。

4.6　练习题

一、填空题

1. 在 Android 中用于显示文字信息的控件是________。
2. 用于显示多条信息列表的控件是________。
3. 让控件隐藏的属性是________，隐藏系统标题栏的方法是________。
4. 引用写好的 xml 布局的关键字是________。

二、选择题

1. 下列属性值中隐藏控件但不占用空间的是(　　)。
   A. gone　　B. visible　　C. visibility　　D. invisible
2. Android 中用于文字大小的单位是(　　),用于控件大小的单位是(　　)。
   A. dp　　B. px　　C. sp　　D. Lp
3. Android 所有控件的主父类是(　　)。
   A. GroupView　　B. TextView　　C. AlertDialog　　D. View

三、简答题

1. 简单描述一下 ListView 的使用步骤。
2. 简单描述一下加载布局的几种方式。

四、编程题

1. 编写一个用户登录界面,要求包含文本框、编辑框和按钮,分别用来显示用户名、密码,输入用户名、密码,登录功能。
2. 设计一个应用,使用 Toast 显示提示信息。

第5章 Intent与组件通信

本章重点

- Intent 启动组件的方法
- 隐式 Intent 及 Intent 相关属性
- Intent 传递数据
- Activity 的启动模式
- 广播消息

Intent 是"意图"的意思,是对将要执行的操作的一种抽象的描述。它可以用来开启一个 Activity,或者将它发送给任何感兴趣的广播接收者 BroadcastReceiver 组件,还可以通过 startService()或者 bindService()与后台的服务 Service 交流。当然,它还可以跨应用交流信息。

Intent 的作用:启动组件并传递数据(putExtra 与 getXxxExtra 方法)。

可见,Intent 与 Android 的四大组件中除 ContentProvider 组件外的其他组件都有关系。

5.1 Intent 概述

Intent 消息对于运行时绑定不同的组件是很方便的,这些组件可以是同一个程序,也可以是不同的程序。一个 Intent 对象是一个被动的数据结构,它保存了一个操作的抽象描述——通常是一个广播的实例、一些发生的事情的描述、一个通知。传递 Intent 到不同组件的机制是互不相同的。

Activity、Services、BroadcastReceiver 是通过 Intent 传递消息的,而另外一个组件 Content Provider 本身就是一种通信机制,不需要通过 Intent。来看下这个图(图 5-1)就知道了。

通过图 5-1 可以看到组件之间通信也是通过 Intent 来完成的。除此之外,两个 Activity 可以把要交换的数据封装成 Bundle 对象,然后通过 Intent 来传递数据。

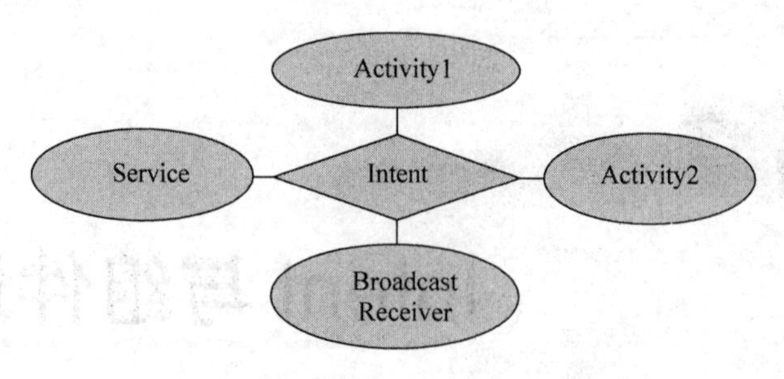

图 5-1　组件之间通信与 Intent 关系

5.2 Intent 启动组件的方法

向 Activity、Service、BroadcastReceiver 这三种组件发送消息的方法，如表 5-1 所示。

表 5-1　Intent 使用的方法

组件名称	方法描述
Activity	startActvity() startActivityForResult()
Service	startService() bindService()
Broadcasts	sendBroadcasts() sendOrderedBroadcasts() sendStickyBroadcasts()

使用 Context. startActivity()或 Activity. startActivityForResult()，通过参数传入一个 Intent 来启动一个 Activity。使用 Activity. setResult()，传入一个 Intent 从 Activity 中返回结果。

将 Intent 对象传给 Context. startService()，可启动一个 Service 或者传消息给一个运行的 Service；将 Intent 对象传给 Context. bindService()，可绑定一个 Service。

将 Intent 对象传给 Context. sendBroadcast()，Context. sendOrderedBroadcast()或者 Context. sendStickyBroadcast()等广播方法，则它们被传给 BroadcastReceiver。

5.3 隐式 Intent 及 Intent 相关属性

Intent 有以下各个组成部分：

- Component(组件)：目的组件；
- Action(动作)：用来表现 Intent 的行动；
- Category(类别)：用来表现动作的类别；
- Data(数据)：表示动作要操纵的数据；
- Type(数据类型)：指定 data 属性的数据类型或 MIME 类型；
- Extras(扩展信息)：扩展信息；

- Flags(标志位)：期望这个意图的运行模式。

Intent 可以分为两类：显式 Intent 和隐式 Intent。

显式 Intent 通过名字指定目标组件，其他程序的开发人员不需要知道组件名。显式 Intent 用于程序内部消息，如 Activity 启动一个下属服务或启动一个姊妹 Activity。

隐式 Intent 没有命名一个目标(组件名是空的)，隐式 Intent 通常用来激活其他程序的组件。

Activity 中 Intent Filter 的匹配过程如图 5-2 所示。

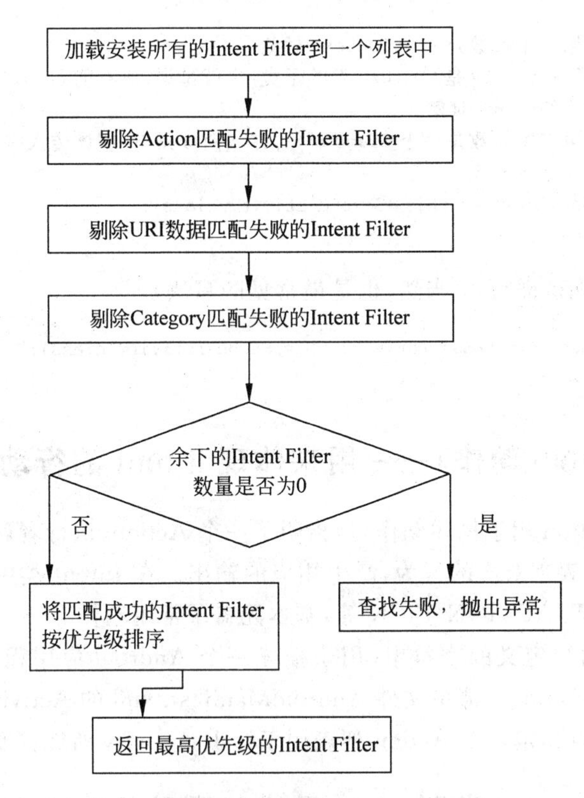

图 5-2 Intent Filter 的匹配过程

隐式 Intent 需要不同的策略。如果没有指定目标，Android 系统需要查找最适合处理 Intent 的组件(或几个组件)——一个单一的 Activity 或服务来执行请求的动作或设置广播接收器来响应广播通知。通过把 Intent 对象的内容和 Intent 管理器比较，判断哪个组件是潜在的接收者。过滤器提供组件的能力并且划定它可以处理的 Intent。它开启可以接收隐式 Intent 的组件类型。如果组件没有 Intent 过滤器，它仅仅可以接收显式的 Intent。含有过滤器的组件既可以接收隐式 Intent，也可以接收显式 Intent。

5.3.1 Component(组件)——目的组件

Intent 的 Component 属性明确指定 ComponentName 对象。

组件名是可选的。如果已设置，Intent 对象分派到目的类的实例；如果不设置，Android 使用 Intent 的其他信息来本地化合适的目标。例如，启动第二个 Activity 时，可以这样来写：

```
//创建一个 Intent 对象
Intent intent = new Intent();
//创建组件,通过组件来响应
ComponentName component = new ComponentName(MainActivity.this, SecondActivity.class);
intent.setComponent(component);
startActivity(intent);
```

如果写得简单一点,监听事件 onClick()方法里可以这样写:

```
//setClass 函数的第一个参数是一个 Context 对象
//Context 是一个类,Activity 是 Context 类的子类,也就是说,所有的 Activity 对象
//都可以向上转型为 Context 对象
//setClass 函数的第二个参数是一个 Class 对象,在当前场景下,应该传入需要被启动
//的 Activity 类的 class 对象
intent.setClass(MainActivity.this, SecondActivity.class);
startActivity(intent);
```

再简单一点,可以这样写:(当然,也是最常见的写法)

```
Intent intent = new Intent(MainActivity.this,SecondActivity.class);
startActivity(intent);
```

5.3.2 Action(动作)——用来体现 Intent 的行动

在 Intent 中,Action 用于描述动作,当指明了一个 Action,执行者就会依照这个动作的指示,接受相关输入,表现对应的行为,产生相应的输出。在 Intent 类中,定义了一批动作,例如 ACTION_VIEW,ACTION_PICK 等,基本涵盖了常用动作。

Action 是一个用户定义的字符串,用于描述一个 Android 应用程序组件,一个 Intent Filter 可以包含多个 Action。清单文件 AndroidManifest.xml 的 Activity 在定义时,可以在其< intent-filter >节点指定一个 Action 列表用于标识 Activity 所能接受的"动作"。

5.3.3 Category(类别)——用来体现动作的类别

Category 属性也是作为< intent-filter >子元素来声明的。例如:

```
<intent-filter>
    <action android:name="com.vince.intent.MY_ACTION"></action>
    <category android:name="com.vince.intent.MY_CATEGORY"></category>
    <category android:name="android.intent.category.DEFAULT"></category>
</intent-filter>
```

Action 和 Category 通常是放在一起用的。例如,新建一个工程文件 test,在默认文件的基础之上,新建文件 SecondActivity.java 和 activity_second.xml。

现在,要到清单文件中进行注册,打开 AndroidManifest.xml,添加 SecondActivity 的 Action 和 Category 的过滤器:

```
<activity android:name=".SecondActivity">
<intent-filter>
    <action android:name="com.qsd.test.MY_ACTION"/>
```

```
            <category android:name = "android.intent.category.DEFAULT" />
        </intent - filter>
    </activity>
```

上面代码表示SecondActivity可以匹配MY_ACTION这个动作，此时，如果在其他的Acticity通过这个Action的条件来查找，那么SecondActivity就具备了这个条件。类似于相亲时，我要求对方有哪些条件，然后对方这个SecondActivity恰巧满足了这个条件。

注：如果没有指定的Category，则必须使用默认的DEFAULT。

也就是说：只有< action >和< category >中的内容同时能够匹配上Intent中指定的Action和Category时，这个活动才能响应Intent。如果使用的是DEFAULT这种默认的Category，在稍后调用startActivity()方法的时候会自动将这个Category添加到Intent中。

现在来修改MainActivity.java中按钮的单击事件，代码如下：

```
button1.setOnClickListener(new OnClickListener() {
            @Override
            public void onClick(View v) {
                //启动另一个Activity,(通过action属性进行查找)
                Intent intent = new Intent();
                //设置动作(实际action属性就是一个字符串标记而已)
//方法: Intent android.content.Intent.setAction(String action)
                intent.setAction("com.qsd.test.MY_ACTION");
               startActivity(intent);
            }
        });
```

上方的代码也可以换成下面这种简洁的方式：

```
button1.setOnClickListener(new OnClickListener() {
            @Override
            public void onClick(View v) {
                //启动另一个Activity,(通过action属性进行查找)
                Intent intent = new Intent("com.qsd.test.MY_ACTION");
                startActivity(intent);
            }
        });
```

上面代码中，在这个Intent中并没有指定具体哪一个Activity，只是指定了一个Action的常量，隐式Intent的作用体现得淋漓尽致。此时，单击MainActivity中的按钮，就会跳到SecondActivity中去。

上述情况只有SecondActicity匹配成功。如果有多个组件匹配成功，就会以对话框列表的方式让用户进行选择。下面做详细介绍。

新建文件ThirdActicity.java和activity_third.xml，然后在清单文件AndroidManifest.xml中添加ThirdActivity的Action和Category的过滤器：

```
<activity android:name = ".ThirdActivity">
    <intent - filter>
        <action android:name = "com.qsd.test.MY_ACTION" />
```

```
        <category android:name = "android.intent.category.DEFAULT" />
    </intent-filter>
</activity>
```

运行程序，当单击 MainActivity 中的按钮时，弹出界面，如图 5-3 所示。

图 5-3 弹出的界面

相信大家看到了这个界面应该就一目了然了。于是可以做出如下总结：

在自定义动作，使用 Activity 组件时，必须添加一个默认的类别，具体的实现为：

```
<intent-filter>
        <action android:name = "com.qsd.test.MY_ACTION"/>
        <category android:name = "android.intent.category.DEFAULT"/>
</intent-filter>
```

如果有多个组件被匹配成功，就会以对话框列表的方式让用户进行选择。每个 Intent 中只能指定一个 Action，但却能指定多个 Category；类别越多，动作越具体，意图越明确。

目前的 Intent 中只有一个默认的 Category，现在可以通过 intent.addCategory()方法来实现。修改 MainActivity 中按钮的单击事件，代码如下：

```
button1.setOnClickListener(new View.OnClickListener() {
@Override
public void onClick(View view) {
    //启动另一个 Activity
    Intent intent = new Intent();
    //设置动作
    intent.setAction("com.qsd.test.MY_ACTION");
```

```
            intent.addCategory("com.qsd.test.MY_CATEGORY");
            startActivity(intent);
        }
    });
```

既然在 Intent 中增加了一个 Category，那么要在清单文件中去声明这个 Category，不然程序将无法运行。代码如下：

```
<activity android:name = ".SecondActivity">
<intent - filter>
    <action android:name = "com.qsd.test.MY_ACTION" />

    <category android:name = "android.intent.category.DEFAULT" />
    <category android:name = "com.qsd.test.MY_CATEGORY" />
</intent - filter>
</activity>
```

此时，单击 MainActicity 中的按钮，就会跳到 SecondActicity 中去。

对自定义类别总结如下：

在 Intent 添加类别可以添加多个类别，要求被匹配的组件必须同时满足这多个类别，才能匹配成功。操作 Activity 的时候，如果没有类别，需加上默认类别。

5.3.4 Data(数据)——表示与动作要操纵的数据

Data 属性是 Android 要访问的数据。和 Action 和 Category 声明方式相同，也是在 <intent-filter>中声明。

多个组件匹配成功显示优先级高的组件，优先级相同的显示组件列表，由用户选择启动。

Data 是用一个 URI 对象来表示的。URI 代表数据的地址，属于一种标识符。通常情况下，使用 action+data 属性的组合来描述一个 Intent——做什么。

使用隐式 Intent，不仅可以启动自己程序内的活动，还可以启动其他程序的活动，这使得 Android 多个应用程序之间的功能共享成为了可能。例如应用程序中需要展示一个网页，没有必要自己去实现一个浏览器(事实上也不太容易)，而是只需要调用系统的浏览器来打开这个网页就行了。

【实例】 打开指定网页：

MainActivity.java 中，监听器部分的核心代码如下：

```
button1.setOnClickListener(new View.OnClickListener() {
@Override
public void onClick(View view) {
    Intent intent = new Intent();
    intent.setAction(Intent.ACTION_VIEW);
    Uri data = Uri.parse("http://www.sqlite.org");
    intent.setData(data);
    startActivity(intent);
    }
});
```

当然，上面的 Intent 代码也可以简写成：

```
Intent intent = new Intent(Intent.ACTION_VIEW);
intent.setData(Uri.parse("http://www.sqlite.org"));
startActivity(intent);
```

代码中指定了 Intent 的 Action 是 Intent.ACTION_VIEW，表示查看的意思，这是一个 Android 系统内置的动作；通过 Uri.parse()方法，将一个网址字符串解析成一个 URI 对象，再调用 Intent 的 setData()方法将这个 URI 对象传递进去。

当单击按钮时，将跳到如图 5-4 所示的界面。

图 5-4 跳转后的界面

5.3.5 Type(数据类型)——对于 data 范例的描写

如果 Intent 对象中既包含 URI 又包含 Type，那么，在< intent-filter >中也必须二者都包含才能通过测试。

Type 属性用于明确指定 Data 属性的数据类型或 MIME 类型，但是通常来说，当 Intent 不指定 Data 属性时，Type 属性才会起作用，否则 Android 系统将会根据 Data 属性值来分析数据的类型，所以无须指定 Type 属性。

Data 和 Type 属性一般只需要一个。通过 setData 方法会把 Type 属性设置为 null，而设置 setType 方法会把 Data 设置为 null，如果想要两个属性同时设置，应使用 Intent.setDataAndType()方法。

【任务】 data+type 属性的使用。

【实例】 播放指定路径的 mp3 文件。

在虚拟机中添加测试文件 libai.mp3。添加步骤是，打开虚拟机之后找到如图 5-5 所示的窗口，之后找到 sdcard 文件夹把文件添加进去。

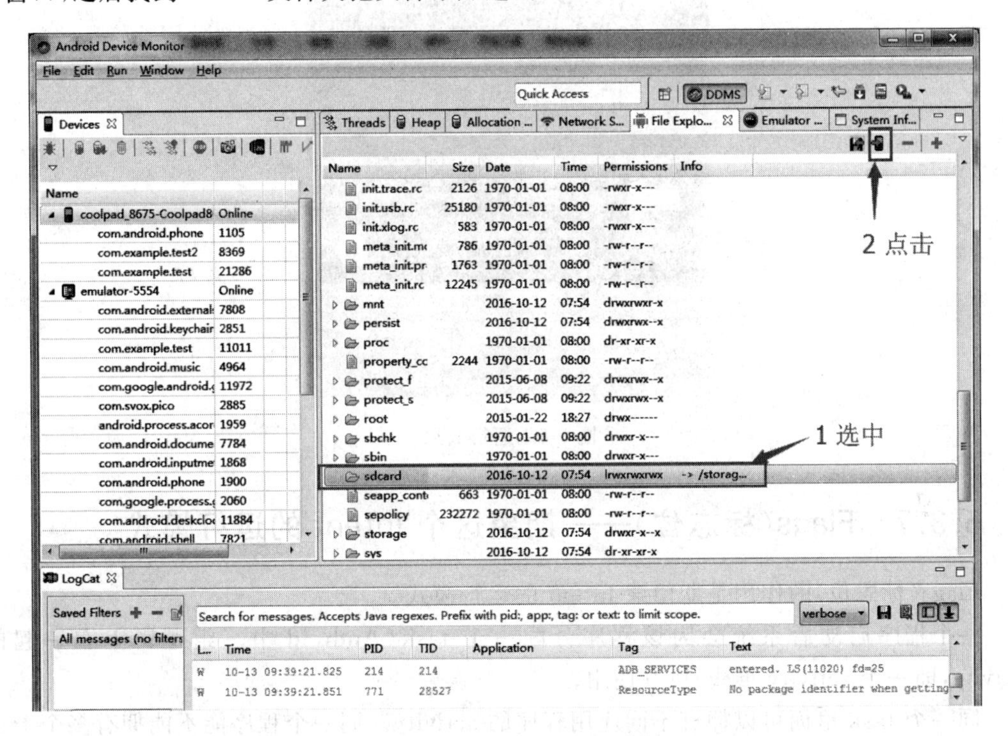

图 5-5 打开虚拟机之后的窗口

修改 MainActivity.java 中按钮监听事件部分的代码如下：

```
button1.setOnClickListener(new View.OnClickListener() {
@Override
public void onClick(View view) {
    Intent intent = new Intent();
    intent.setAction(Intent.ACTION_VIEW);
   //获得文件路径,后边是虚拟机存储卡的路径
    Uri data = Uri.fromFile(new File("/mnt/sdcard/libai.mp3"));
    //设置 data + type 属性
//方法: Intent android.content.Intent.setDataAndType(Uri data, String type)
intent.setDataAndType(data, "audio/ * ");
startActivity(intent);
}
});
```

运行后，当单击按钮时，效果如图 5-6 所示。

5.3.6 Extras(扩展信息)——扩展信息

使用 Extras 可以为组件提供扩展信息，是其他所有附加信息的集合。例如，如果要执行“发送电子邮件”这个动作，可以将电子邮件的标题、正文等保存在 Extras 里，传给电子邮件发送组件。

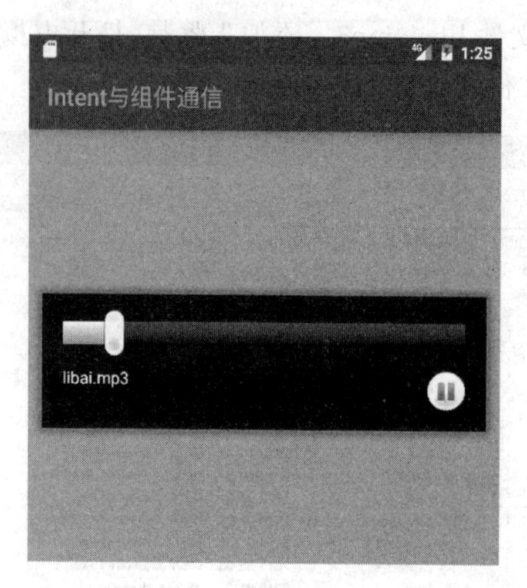

图 5-6 运行效果

5.3.7 Flags(标志位)——期望这个 Intent 的运行模式

Flags(标志位)的作用是期望该 Intent 的运行模式。

一个程序启动后系统会为该程序分配一个 task(Android 中一组逻辑上在一起的 activity,即一个 activity 堆栈)供其使用。

同一个 task 里面可以拥有不同应用程序的 activity。同一个程序能不能拥有多个 task 与加载 activity 的启动模式有关。

5.4 更多隐式 Intent

5.4.1 打开指定网页

在 MainActivity.java 中,监听器部分的核心代码如下(之前的代码):

```
button1.setOnClickListener(new OnClickListener() {
        @Override
        public void onClick(View v) {
            Intent intent = new Intent();
            intent.setAction(Intent.ACTION_VIEW);
            Uri data = Uri.parse("http://www.baidu.com");
            intent.setData(data);
            startActivity(intent);
        }
});
```

当然,这段代码也可以简写成:

```
button1.setOnClickListener(new OnClickListener() {
        @Override
        public void onClick(View v) {
```

```
                Intent intent = new Intent(Intent.ACTION_VIEW);
                intent.setData(Uri.parse("http://www.baidu.com"));
                startActivity(intent);
            }
        });
```

或者可以写成：

```
button1.setOnClickListener(new OnClickListener() {
            @Override
            public void onClick(View v) {
                Uri uri = Uri.parse("http://www.baidu.com");
                Intent intent = new Intent(Intent.ACTION_VIEW,uri);
                startActivity(intent);
            }
        });
```

代码中指定了 Intent 的 Action 是 Intent.ACTION_VIEW，表示查看的意思，这是一个 Android 系统内置的动作；通过 Uri.parse()方法，可将一个网址字符串解析成一个 URI 对象，再调用 Intent 的 setData()方法将这个 URI 对象传递进去。

5.4.2　打电话

【方式一】 打开拨打电话的界面：

```
Intent intent = new Intent(Intent.ACTION_DIAL);
intent.setData(Uri.parse("tel:10086"));
startActivity(intent);
```

运行程序后，单击"电话"按钮，效果如图 5-7 所示。

图 5-7　拨打电话界面

【方式二】 直接拨打电话：

```
Intent intent = new Intent(Intent.ACTION_CALL);
intent.setData(Uri.parse("tel:10086"));
startActivity(intent);
```

要使用这个功能必须在清单文件中加入权限(加一行代码)：

```
<uses-sdk android:minSdkVersion="8" android:targetSdkVersion="16" />
<uses-permission android:name="android.permission.CALL_PHONE"/>
```

5.4.3 发送短信

【方式一】 打开发送短信的界面。

```
Intent intent = new Intent(Intent.ACTION_VIEW);
intent.setType("vnd.android-dir/mms-sms");
//"sms_body"为固定内容
intent.putExtra("sms_body", "具体短信内容");
startActivity(intent);
```

【方式二】 打开发送短信的界面(同时指定电话号码)。

```
Intent intent = new Intent(Intent.ACTION_SENDTO);
intent.setData(Uri.parse("smsto:18780260012"));
//"sms_body"为固定内容
intent.putExtra("sms_body", "具体短信内容");
startActivity(intent);
```

5.4.4 播放指定路径音乐

```
    Intent intent = new Intent();
    intent.setAction(Intent.ACTION_VIEW);
  //获得文件路径,后边是虚拟机存储卡的路径
    Uri data = Uri.fromFile(new File("/mnt/sdcard/libai.mp3"));
    //设置 data + type 属性
//方法: Intent android.content.Intent.setDataAndType(Uri data, String type)
intent.setDataAndType(data, "audio/*");
startActivity(intent);
```

5.4.5 卸载程序

注：无论是安装还是卸载,应用程序都是根据包名 package 来识别的。

```
Intent intent = new Intent(Intent.ACTION_DELETE);
Uri data = Uri.parse("package:com.qsd.ch5_3");
intent.setData(data);
startActivity(intent);
```

5.4.6 安装程序

安装程序的前提是有这个文件,如果没有会报错。安装前应进行文件是否存在的判断。

本例的这个文件是已提前导入了的。

```
Intent intent = new Intent(Intent.ACTION_VIEW);
        // 路径不能写成: "file:///storage/sdcard0/···"
        File file = new File("/storage/sdcard/ch5_4.apk");
        Uri data = Uri.fromFile(file);
        // Type 的字符串为固定内容
        intent.setDataAndType(data,
                "application/vnd.android.package-archive");
        startActivity(intent);
```

路径不能写成："file:///storage/sdcard0/…"，不然报错，如图5-8所示。

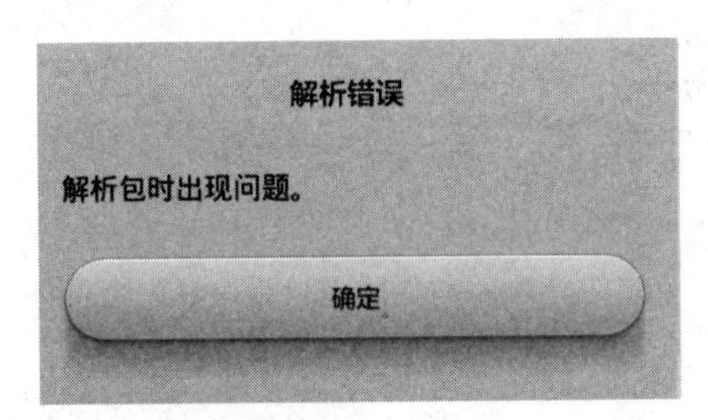

图5-8　因路径错误导致的报错

问题：通过下面的这种方式安装程序，运行时为什么会出错呢？

```
//通过指定的 action 来安装程序
Intent intent = new Intent(Intent.ACTION_PACKAGE_ADDED);

Uri data = Uri.fromFile(new File("/storage/sdcard/ch5_4.apk"));
intent.setData(data);
startActivity(intent);
```

5.5 传递数据

5.5.1 显式Intent

前面的隐式Intent操作通常都是打开一些系统的活动，下面介绍一下显性Intent的操作。

新建一个activity_first.xml布局文件，代码如下：

```
<RelativeLayout xmlns:android="http://schemas.android.com/apk/res/android"
xmlns:tools="http://schemas.android.com/tools"
android:layout_width="match_parent"
android:layout_height="match_parent"
>

<Button
    android:id="@+id/btn_close"
    android:layout_width="wrap_content"
```

```
        android:layout_height = "wrap_content"
        android:padding = "10dp"
        android:text = "关闭" />
</RelativeLayout>
```

代码中添加一个“关闭”按钮，按钮显示的是关闭。然后再新建一个 FirstActivity，主要代码如下：

```
package com.qsd;

import android.app.Activity;
import android.content.Intent;
import android.os.Bundle;
import android.view.View;
import android.widget.Button;

import com.qsd.ch5_5.R;

public class FirstActivity extends Activity {
    @Override
    protected void onCreate(Bundle savedInstanceState) {
        super.onCreate(savedInstanceState);
        setContentView(R.layout.activity_first);
        Button button = (Button) findViewById(R.id.btn_close);
        button.setOnClickListener(new View.OnClickListener() {
            @Override
            public void onClick(View view) {
                FirstActivity.this.finish();
            }
        });
    }
}
```

代码中实例化了 Button 并添加单击事件，单击事件里面多了一个 finish()方法，此方法的作用是关闭当前的活动。之后在 AndroidManifest.xml 中为新建的 FirstActivity 进行注册。

```
<manifest xmlns:android = "http://schemas.android.com/apk/res/android"
    package = "com.qsd.ch5_5"
    android:versionCode = "1"
    android:versionName = "1.0" >

    <uses - sdk
        android:minSdkVersion = "8"
        android:targetSdkVersion = "21" />

    <application
        android:allowBackup = "true"
        android:icon = "@drawable/ic_launcher"
        android:label = "@string/app_name"
        android:theme = "@style/android:Theme.Holo.Light" >
```

```
        <activity android:name = "com.qsd.MainActivity" >
            <intent - filter >
                <action android:name = "android.intent.action.MAIN" />

                <category android:name = "android.intent.category.LAUNCHER" />
            </intent - filter >
        </activity >
        <activity android:name = "com.qsd.FirstActivity" />
    </application >

</manifest >
```

对 FirstActivity 只做了简单的注册，但并不影响使用。下面来修改 activity_main.xml 中的代码。

```
<RelativeLayout xmlns:android = "http://schemas.android.com/apk/res/android"
    xmlns:tools = "http://schemas.android.com/tools"
    android:layout_width = "match_parent"
    android:layout_height = "match_parent"
  >

    <Button
        android:id = "@ + id/btn_link"
        android:layout_width = "wrap_content"
        android:layout_height = "wrap_content"
        android:text = "单击跳转" />
</RelativeLayout >
```

这里添加了一个"单击跳转"的 Button 按钮。再修改 MainActivity 中的主要代码。

```
package com.qsd;

    import android.app.Activity;
    import android.content.Intent;
    import android.os.Bundle;
    import android.view.View;
    import android.widget.Button;

    import com.qsd.ch5_5.R;

    public class MainActivity extends Activity {
        @Override
        protected void onCreate(Bundle savedInstanceState) {
            // TODO Auto - generated method stub
            super.onCreate(savedInstanceState);
            setContentView(R.layout.activity_main);
            Button button = (Button) findViewById(R.id.btn_link);
            button.setOnClickListener(new View.OnClickListener() {
                @Override
                public void onClick(View view) {
                    Intent intent = new Intent(MainActivity.this,
```

```
                        FirstActivity.class);
                startActivity(intent);
            }
        });

    }
}
```

图 5-9　启动 Activity 的效果

上面的代码也是实例化一个按钮再添加单击事件。事件中用到了 Intent 的构造方法，第一个参数 Context 要提供启动活动的上下文，第二个参数 Class 则是指定想要启动的目标活动，通过这个 startActivity()方法启动活动。效果如图 5-9 所示。

可以看到这样也是可以启动 Activity 的，如果想返回上一个 Activity，只需要单击"关闭"按钮就行，单击手机的 Back 键也是一样的。使用显式 Intent 是要明确地告诉系统要启动谁。

5.5.2　向下一个活动传递数据

在 Activity 中启动另一个 Activity 时，常常需要把一些数据传递过去。前面介绍了怎么启动活动，现在介绍一下 Intent 中的数据传递。

在隐式 Intent 启动活动中介绍过不少其中的方法和属性，这一节介绍常用的 putExtra()的一系列方法重载，这些方法可以把自己想要的数据暂存 Intent 中，启动另一个 Activity 后，只需要把这些数据从 Intent 中取出来即可。下面修改 MainActivity 中的方法，代码如下：

```
@Override
protected void onCreate(Bundle savedInstanceState) {
    super.onCreate(savedInstanceState);
    setContentView(R.layout.activity_main);
    Button button = (Button) findViewById(R.id.btn_link);
    button.setOnClickListener(new View.OnClickListener() {
        @Override
        public void onClick(View view) {
          Intent intent = new Intent(MainActivity.this, FirstActivity.class);
            intent.putExtra("data","传过来的一条数据");
            startActivity(intent);
        }
    });
}
```

这里通过 putExtra()方法传过去了一串字符。这个方法接收两个参数，第一个是参数的键值，第二个是参数的值，也就是传过去的数据。

然后再修改 FirstActivity 中的代码并显示出来，代码如下所示：

```
@Override
```

```
protected void onCreate(Bundle savedInstanceState) {
    super.onCreate(savedInstanceState);
    setContentView(R.layout.activity_first);
    Button button = (Button) findViewById(R.id.btn_close);

    Intent intent = getIntent();
    String data = intent.getStringExtra("data");
    button.setText(data);
}
```

这里用到了 getIntent()方法来获取 Intent 对象，然后通过调用 getStringExtra()方法传入相对应的键值，这样就得到先前传递的数据了。由于传过来的数据是字符串类型的，所以接收的时候要用 getStringExtra()方法来获取数据。如果传过来的是整型数据就要用 getIntExtra()方法；如果传过来的数据是布尔类型就用 getBooleanExtra()方法；如果传过来的是字符串数组就用 getStringArrayExtra()方法，等等。获取数据之后，把数据显示到 Button 按钮上，效果如图 5-10 所示。

图 5-10 用 Button 按钮显示传递过来的数据

5.5.3 返回数据给上一个活动

既然可以传递数据给下一个 Activity，那么能不能传递数据回去呢？答案是肯定的。返回上一个活动只需关闭当前的活动即可，不需要再启动一个活动来传递数据。之前启动活动的时候介绍过两种方法：一种是 startActivity()方法，这也是一直在用的方法；另一种是 startActivityForResult()方法，这个方法是在活动销毁的时候返回一些数据给上一个活动。

修改 MainActivity 中的启动方法，代码如下所示：

```
@Override
protected void onCreate(Bundle savedInstanceState) {
    super.onCreate(savedInstanceState);
    setContentView(R.layout.activity_main);
    button = (Button) findViewById(R.id.btn_link);
    button.setOnClickListener(new View.OnClickListener() {
        @Override
        public void onClick(View view) {
            Intent intent = new Intent(MainActivity.this, FirstActivity.class);
            intent.putExtra("data", "传过来的一条数据");
            startActivityForResult(intent, 1);
        }
    });
}
```

startActivityForResult()方法中接收两个参数，第一个是Intent，第二个是请求码。请求码可以随意设置，但是必须是自然数，这里传入的是1。请求码主要用于区别返回的数据。

在FirstActivity中修改单击事件，代码如下：

```
@Override
protected void onCreate(Bundle savedInstanceState) {
    super.onCreate(savedInstanceState);
    setContentView(R.layout.activity_first);
    Button button = (Button) findViewById(R.id.btn_close);

    Intent intent = getIntent();
    String data = intent.getStringExtra("data");
    button.setText(data);
    button.setOnClickListener(new View.OnClickListener() {
        @Override
        public void onClick(View view) {
            Intent intent1 = new Intent();
            intent1.putExtra("data_return", "这是返回的数据");
            setResult(RESULT_OK, intent1);
            finish();
        }
    });
}
```

在此基础上，添加Intent对象并传入一个键值为data_return的数据，用setResult方法向上一个Activity中返回键值为RESULT_OK的Intent，然后调用finish()方法结束当前的Activity。

由于这里使用的是startActivityForResult()方法启动的，如果想要得到被销毁之后的Activity的数据，那就需要在MainActivity中重写onActivityResult()方法。代码如下所示：

```
@Override
protected void onActivityResult(int requestCode, int resultCode, Intent data) {
    super.onActivityResult(requestCode, resultCode, data);
    if (data != null) {
        switch (requestCode) {
            case 1:
                if (resultCode == RESULT_OK) {
                    String returnData = data.getStringExtra("data_return");
                    button.setText(returnData);
                }
        }
    }
}
```

onActivityResult()方法中有三个参数，第一个参数requestCode，就是在启动Activity的时候传入的请求码；第二个参数resultCode是返回数据时传入的处理结果；第三个参数是data，这个就是返回的数据。onActivityResult()方法中先判断data是否为空，如果确定不为空之后再判断 resultCode 的值。由于启动的 Activity 比较多，所以要通过 requestCode 的值来判断数据来源，这个值主要由 RESULT_OK 和 RESULT_CANCELED 来判断返回数据是否成功。如果都符合要求，那么就从 data 中获取值并显示到按钮中。效果如图 5-11、图 5-12 所示。

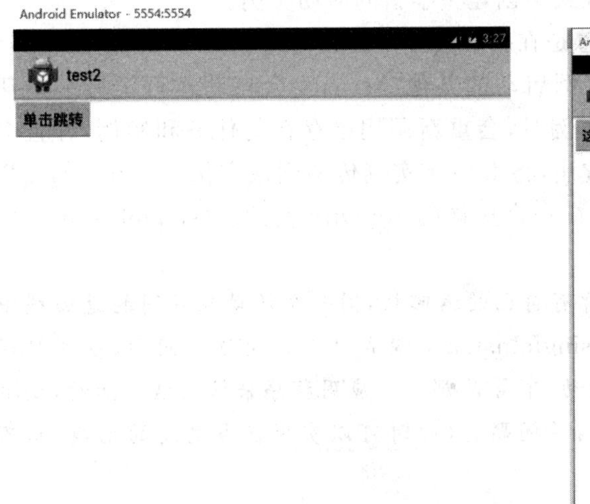

图 5-11　原来的数据

图 5-12　返回的数据

手机可以用 Back 键回到上一个 Activity 中，道理其实和 finish()是一样的。如果想要与上面效果图一样的效果的话，就要重写 onBackPressed()方法。代码如下所示：

```
    @Override
public void onBackPressed() {
    Intent intent1 = new Intent();
    intent1.putExtra("data_return", "这是返回的数据");
    setResult(RESULT_OK, intent1);
    finish();
}
```

5.6　Activity 的启动模式

Activity 有四种启动模式：standard、singleTop、singleTask、singleInstance。可以在清单文件 AndroidManifest.xml 中 activity 标签的属性 android:launchMode 中设置。

- standard 模式：这是默认的模式。以这种模式加载时，每当启动一个新的 Activity，必定会构造一个新的 Activity 实例放到返回栈（目标 task）的栈顶，而不管这个 Activity 是否已经存在于返回栈中。
- singleTop 模式：如果一个以 singleTop 模式启动的 Activity 的实例已经存在于返回栈的栈顶，那么再启动这个 Activity 时，不会创建新的实例，而是重用位于栈顶的这个实例，并且会调用该实例的 onNewIntent()方法将 Intent 对象传递到这个实例中。

注：如果以 singleTop 模式启动的 Activity 的一个实例已经存在于返回栈中，但是不在栈顶，那么它的行为和 standard 模式相同，也会创建多个实例。

- singleTask 模式：在这种模式下，每次启动一个 Activity 时，系统首先会在返回栈中检查是否存在该活动的实例，如果存在，则直接使用该实例，并把这个活动之上的所有活动统统清除；如果没有发现就会创建一个新的活动实例。
- singleInstance 模式：这种模式总是在新的任务中开启，并且这个新的任务中有且只有这一个实例，也就是说被该实例启动的其他 Activity 会自动运行于另一个任务中。当再次启动该 Activity 的实例时，会重新调用已存在的任务和实例，并且会调用这个实例的 onNewIntent()方法，将 Intent 实例传递到该实例中。和 singleTask 相同，同一时刻在系统中只会存在一个这样的 Activity 实例。（singleInstance 即单实例。）

注：前三种模式中，每个应用程序都有自己的返回栈，同一个活动在不同的返回栈中入栈时，必然是创建了新的实例。而使用 singleInstance 模式可以解决这个问题，在这种模式下会有一个单独的返回栈来管理这个活动，不管是哪一个应用程序来访问这个活动，都共用同一个返回栈，也就解决了共享活动实例的问题。（此时可以实现任务之间的切换，而不是单独某个栈中的实例切换。）

其实，不在清单文件中设置，而在代码中通过 Flag 来设置也是可以的，如下：

```
Intent intent = new Intent(MainActivity.this,SecondActivity.class);
//相当于 singleTask
intent.setFlags(Intent.FLAG_ACTIVITY_NEW_TASK);
startActivity(intent);

Intent intent = new Intent(MainActivity.this,SecondActivity.class);
//相当于 singleTop
intent.setFlags(Intent.FLAG_ACTIVITY_CLEAR_TOP);
startActivity(intent);
```

5.7 广播消息

5.7.1 BroadcastReceiver 简介

BroadcastReceiver 直译是“广播接收者”，所以它的作用是接收发送过来的广播，那么什么是广播呢？广播，可以理解成系统中消息的一种变种，就是当一个事件发生时，例如系统突然断网，系统就发一个广播消息给所有的接收者，所有的接收者在得到这个消息之后，就知道现在没网络了。

5.7.2 发送广播

在程序中使用广播很简单，与之前用隐式的 Intent 打开指定网页等操作类似。第一步新建一个 MyReceiver 让它继承 BroadcastReceiver 并实现里面的 onReceive()方法，代码如下所示：

```
package com.qsd.ch5_7_2;

import android.content.BroadcastReceiver;
import android.content.Context;
import android.content.Intent;
import android.util.Log;

public class MyReceiver extends BroadcastReceiver {

    @Override
    public void onReceive(Context context, Intent intent) {
        String msg = intent.getStringExtra("msg");
        Log.e("MyReceiver", "接收的消息内容是: " + msg);

    }

}
```

在这段代码中我们获取了键值为 msg 的字符串数据并显示出来。

第二步在 AndroidManifest.xml 中配置一下 Action，代码如下所示：

```
<receiver android:name = "com.qsd.ch5_7_2.MyReceiver" >
        <intent - filter >

          //指定该 BroadcastReceiver 所响应的 Intent 的 Action
            <action android:name = "com.qsd.MY_RECEIVER" >
            </action >
        </intent - filter >
    </receiver >
```

第三步修改 MainActivity 中代码，添加一个 Button 并给它添加单击事件，再添加一个 MyReceiver1 类(代码和 MyReceiver 类一样)用于测试，代码如下所示：

```
package com.qsd.ch5_7_2;

import android.app.Activity;
import android.content.Intent;
import android.os.Bundle;
import android.view.View;
import android.widget.Button;

import com.qsd.ch5_7.R;

public class MainActivity extends Activity {
    @Override
    protected void onCreate(Bundle savedInstanceState) {
        // TODO Auto - generated method stub
        super.onCreate(savedInstanceState);
        setContentView(R.layout.activity_main);
        Button button = (Button) findViewById(R.id.btn_link);
        button.setOnClickListener(new View.OnClickListener() {
```

```
            @Override
            public void onClick(View view) {
                // 创建 Intent 对象
                Intent intent = new Intent();
                // 设置 Intent 的 Action 属性
                intent.setAction("com.qsd.MY_RECEIVER");
                // 保存数据到 Intent 中
                intent.putExtra("msg", "这是一条简单的广播消息");
                // 发送广播
                sendBroadcast(intent);
            }
        });

    }
}
```

程序很简单，只是在最后一行中添加了 sendBroadcast()方法用于发送广播。运行之后效果如图 5-13 所示。

com.qsd.ch5_7	MyReceiver	接收的消息内容是：这是一条简单的广播消息
com.qsd.ch5_7	MyReceiver1	接收的消息内容是：这是一条简单的广播消息

图 5-13 广播消息

5.7.3 发送有序广播

普通广播是指大家等级都是一样的，当广播到来时，都能接收到，并没有接收的先后顺序。由于是同时接收到的，所以一个接收者是没有办法阻止另一个接收者接收这个广播的。

有序广播是指接收者是按一定的优先级顺序来接收的，优先级高的先收到，并可以对广播进行操作后，再传给下一个接收者。当然也可以不传。如果不传的话，后面的接收者就都收不到这个广播了。

先前例子的代码就是普通的广播，下面介绍有序广播。修改 MainActivity 中的程序，代码如下：

```
package com.qsd.ch5_7_3;

import android.app.Activity;
import android.content.Intent;
import android.os.Bundle;
import android.view.View;
import android.widget.Button;

import com.qsd.ch5_7.R;

public class MainActivity extends Activity {
    @Override
    protected void onCreate(Bundle savedInstanceState) {
        // TODO Auto-generated method stub
        super.onCreate(savedInstanceState);
        setContentView(R.layout.activity_main);
```

```
        Button button = (Button) findViewById(R.id.btn_link);
        button.setOnClickListener(new View.OnClickListener() {
            @Override
            public void onClick(View view) {
                // 创建一个 Intent 对象
                Intent intent = new Intent();
                // 设置 Intent 的 Action 属性
                intent.setAction("com.qsd.MY_RECEIVER");
                // 保存数据到 Intent 中
                intent.putExtra("msg", "这是一条简单的广播消息");
                // 发送有序广播
                sendOrderedBroadcast(intent, null);
            }
        });

    }
}
```

代码中只修改发送广播的 sendOrderedBroadcast()方法，此方法的第一个参数是 Intent，第二个参数是一个与权限相关的字符串，这里设置为 null。运行之后大家会看到和普通广播是一样的，看上去没有区别，但是有序广播允许截断接收，只需写一个方法就行，代码如下所示：

```
package com.qsd.ch5_7_3;

import android.content.BroadcastReceiver;
import android.content.Context;
import android.content.Intent;
import android.util.Log;

public class MyReceiver extends BroadcastReceiver {

    @Override
    public void onReceive(Context context, Intent intent) {
        String msg = intent.getStringExtra("msg");
        Log.e("MyReceiver", "接收的消息内容是: " + msg);
        // 取消 Broadcast 的继续传播
        abortBroadcast();
    }

}
```

我们只是在代码中添加了 abortBroadcast()方法，表示将这条广播截断，后面的广播接收者将无法收到这条广播了。重新运行看一下效果，会发现只有一条了。

5.7.4 接收系统广播

除了接收用户发送的广播之外，还可以接收一些系统的广播。这非常实用。有很多我们经常会遇到的应用场合，例如监听用户来电、监听用户短信、拦截黑名单电话以及需要实现开机启动消息推送服务，等等。

新建一个 NetReceiver 类，代码如下所示：

```
package com.qsd.ch5_7_4;

import android.content.BroadcastReceiver;
import android.content.Context;
import android.content.Intent;
import android.net.ConnectivityManager;
import android.net.NetworkInfo;
import android.util.Log;

public class NetReceiver extends BroadcastReceiver {

    @Override
    public void onReceive(final Context context, Intent intent) {
        if (isNetworkAvailable(context)) {
            Log.e("Net", "有网了");
        } else {
            Log.e("Net", "没网了");
        }
    }

    /**
     * 网络是否可用
     *
     * @param context
     * @return
     */
    public static boolean isNetworkAvailable(Context context) {
        ConnectivityManager mgr = (ConnectivityManager) context
                .getSystemService(Context.CONNECTIVITY_SERVICE);
        NetworkInfo[] info = mgr.getAllNetworkInfo();
        if (info != null) {
            for (int i = 0; i < info.length; i++) {
                if (info[i].getState() == NetworkInfo.State.CONNECTED) {
                    return true;
                }
            }
        }
        return false;
    }
}
```

在清单文件 AndroidManifest.xml 中添加：

```
<receiver android:name = "com.qsd.ch5_7_4.NetReceiver" >
        <intent-filter>
            <action android:name = "android.net.conn.CONNECTIVITY_CHANGE" >
            </action>
        </intent-filter>
    </receiver>
```

这里用到了系统的 Action 中的 CONNECTIVITY_CHANGE。

在 application 同级目录下添加以下代码。此代码为用户获取网络权限。如果不添加，在打开网络的时候将无法测试。

```
<uses-permission
android:name="android.permission.ACCESS_NETWORK_STATE"></uses-permission>
```

运行代码测试效果如图5-14所示。

com.qsd.ch5_7	Net	没网了
com.qsd.ch5_7	Net	有网了

图5-14

5.8 本章小结

本章主要介绍了Intent与组件的通信，Intent的Component、Action、Category、Data、Type各种属性及其用法，如何在清单文件AndroidMainfest.xml文件中添加<intent-filer>节点，BroadcastReceiver的创建以及如何配置BroadcastReceiver组件，还介绍了在程序中发送Broadcast的方法。

5.9 练习题

一、填空题

1. 在Android中使用Intent启动不同活动的方法有________、________。
2. 在Intent中传递字符串数据的方法是________。
3. 使用Intent启动普通广播的方法是________，启动有序广播的是________，在有序广播中结束传递到下一个接收广播中数据的方法是________。

二、选择题

1. 在AndroidManifest.xml文件中注册BroadcastReceiver方式正确的是(　　)。

A.
```
<receiver android:name="NewBroad">
    <intent-filter>
        <action
    android:name="android.provider.action.NewBroad"/>
<action>
    </intent-filter>
</receiver>
```

B.
```
<receiver android:name="NewBroad">
    <intent-filter>
    android:name="android.provider.action.NewBroad"/>
    </intent-filter>
</receiver>
```

C.
```
<receiver android:name = "NewBroad">
        <action
    android:name = "android.provider.action.NewBroad"/>
<action>
</receiver>
```

D.
```
<intent-filter>
<receiver android:name = "NewBroad">
            <action>
    android:name = "android.provider.action.NewBroad"/>
  <action>
      </receiver>
</intent-filter>
```

2. 不能正确退出 Activity 的方法是(　　)。

A. finish()　　　　B. 抛出异常强制退出

C. System.exit()　　　　D. onStop()

三、简答题

1. Activity 启动模式有哪几种? 它们的区别是什么?
2. 注册广播有几种方式,这些方式各有何优缺点?
3. Intent 可以传递哪些类型的数据?

第6章 Android 后台服务

本章重点

- 服务的创建
- 服务的生命周期
- 服务的启动方式
- 服务通信

6.1 Service 简介

Service 是 Android 中的一个类，它是 Android 的四大组件之一。使用 Service 可以在后台执行长时间的操作。Service 并不与用户产生 UI 交互。其他的应用组件可以启动 Service。即便用户切换到了其他应用，启动的 Service 仍可在后台运行。一个组件可以与 Service 绑定并与之交互，甚至跨进程通信（IPC）。例如，一个 Service 可以在后台执行网络请求、播放音乐、执行文件读写操作或者与 content provider 交互等。

6.2 Service 的基本用法

Service 组件也是可执行的程序，它也有自己的生命周期。创建、配置 Service 的过程与创建、配置 Activity 的过程基本相似。

6.2.1 创建、配置 Service

因为 Service 和 Activity 相似，所以创建 Service 也需要两个步骤：

第一步：定义一个继承 Service 的子类；

第二步：在清单文件 AndroidManifest.xml 中配置该 Service。

创建 MyService 类让它继承 Service，代码如下所示：

```
package com.qsd.ch6_2_1;

import android.app.Service;
import android.content.Intent;
import android.os.IBinder;
import android.util.Log;
```

```
public class MyService extends Service {

    @Override
    public IBinder onBind(Intent intent) {
        return null;
    }

    @Override
    public void onCreate() {
        super.onCreate();
        Log.d("MyService", "onCreate()");
    }

    @Override
    public int onStartCommand(Intent intent, int flags, int startId) {

        Log.d("MyService", "onStartCommand()");
        return super.onStartCommand(intent, flags, startId);
    }

    @Override
    public void onDestroy() {
        super.onDestroy();
        Log.d("MyService", "onDestroy()");
    }
}
```

在上边的代码中，我们重写了onBind()、onCreate()、onStartCommand()、onDestroy()方法。其中，onBind()方法是必须实现的，onCreate()，onStartCommand()，onDestroy()这三个方法是常用的三个方法。定义完Service之后，接下来在清单文件AndroidManifest.xml中配置Service，代码如下所示：

```
<service android:name = "com.qsd.ch6_1.MyService" >
    </service>
```

代码很简单，与配置Activity非常相似，只是配置Service是<service.../>标签。也可以配置<intent-filer.../>子标签，用于说明这个Service被哪些Intent启动。当这些步骤都操作完成之后就可以运行Service了。

6.2.2 启动Service

创建Service后，如何启动它呢？

在MainActivity和相对应的activity_main.xml中修改代码如下(MainActivity.Java的主要部分)：

```
package com.qsd.ch6_2_2;

import android.app.Activity;
```

```
import android.content.Intent;
import android.os.Bundle;
import android.view.View;
import android.widget.Button;

import com.qsd.ch6_2.R;
import com.qsd.ch6_2_1.MyService;

public class MainActivity extends Activity {
    @Override
    protected void onCreate(Bundle savedInstanceState) {
        // TODO Auto-generated method stub
        super.onCreate(savedInstanceState);
        setContentView(R.layout.activity_main);
        //获取界面中的按钮
        Button start = (Button) findViewById(R.id.btn_start);
        Button stop = (Button) findViewById(R.id.btn_stop);
        //创建启动 Service 的 Intent
        final Intent intent = new Intent(this, MyService.class);
        start.setOnClickListener(new View.OnClickListener() {
            @Override
            public void onClick(View view) {
                //启动 Service
                startService(intent);
            }
        });
        stop.setOnClickListener(new View.OnClickListener() {
            @Override
            public void onClick(View view) {
                //关闭 Service
                stopService(intent);
            }
        });

    }
}
```

activity_main.xml 主要部分：

```
<?xml version="1.0" encoding="utf-8"?>
<LinearLayout xmlns:android="http://schemas.android.com/apk/res/android"
    xmlns:tools="http://schemas.android.com/tools"
    android:layout_width="match_parent"
    android:layout_height="match_parent">

    <Button
        android:id="@+id/btn_start"
        android:layout_width="wrap_content"
        android:layout_height="wrap_content"
        android:text="启动 Service" />
```

```
    <Button
        android:id = "@ + id/btn_stop"
        android:layout_width = "wrap_content"
        android:layout_height = "wrap_content"
        android:text = "关闭 Service" />

</LinearLayout>
```

从代码可以看出，启动和关闭非常简单，直接调用 Context 中的 startService()和 stopService()方法就可以启动和关闭 Service。运行之后效果如图 6-1 所示。

当单击“开始”按钮之后就可以看到应用 test2 已经运行了；当单击“停止”按钮之后，应用 test2 就在应用列表消失了，如图 6-2 所示。

图 6-1　可以看到已经运行的应用 test2

图 6-2　应用 test2 在应用列表消失

6.2.3　Service 和 Activity 通信

如果只是想要启动一个后台服务长期运行某项任务，那么可以使用 startService()启动服务。如果想要与正在运行的 Service 取得联系，那么应该使用 bindService()和 unbingdService()方法启动和关闭 Service。

新建一个 MyBindService，让它继承 Service，代码如下：

```
public class MyBindService extends Service {
private int count;
private boolean quit;
// 定义 onBinder 方法所返回的对象
private MyBinder binder = new MyBinder();
```

```
    // 通过继承 Binder 来实现 IBinder 类
    public class MyBinder extends Binder
    {
        public int getCount() {
            // 获取 Service 的运行状态: count
            return count;
        }
    }

    // 必须实现的方法,绑定该 Service 时回调该方法
    @Override
    public IBinder onBind(Intent intent) {
        Log.e("MyBindService", "onBind()");
        // 返回 IBinder 对象
        return binder;
    }

    // Service 被创建时回调该方法
    @Override
    public void onCreate() {
        super.onCreate();
        Log.e("MyBindService", "Created()");
        // 启动一个线程,动态修改 count 状态值
        new Thread() {
            @Override
            public void run() {
                while (!quit) {
                    try {
                        Thread.sleep(1000);
                    } catch (InterruptedException e) {
                    }
                    count++;
                }
            }
        }.start();
    }

    // Service 被断开连接时回调该方法
    @Override
    public boolean onUnbind(Intent intent) {
        Log.e("MyBindService", "onUnbind()");
        return true;
    }

    // Service 被关闭之前回调该方法
    @Override
    public void onDestroy() {
        super.onDestroy();
        this.quit = true;
        Log.e("MyBindService", "onDestroy()");
    }
}
```

在之前的代码中，在onBind()方法中并没有返回信息，在实际的开发过程中通常会采用继承Binder的方式实现自己的IBinder对象。下面代码很明确地实现了onBind()方法，并让该方法返回一个有效的IBinder对象。在下面代码中开了一个小线程用于做数据累加测试，再修改MainActivity代码，如下所示：

```
public class MainActivity extends Activity {

// 保持所启动的 Service 的 IBinder 对象
TextView mtvCount;
MyBindService.MyBinder binder;
// 定义一个 ServiceConnection 对象
private ServiceConnection conn = new ServiceConnection() {
    // 当该 Activity 与 Service 连接成功时回调该方法
    @Override
    public void onServiceConnected(ComponentName name, IBinder service) {
        // 获取 Service 的 onBind 方法所返回的 MyBinder 对象
        binder = (MyBindService.MyBinder) service;
    }

    // 当该 Activity 与 Service 断开连接时回调该方法
    @Override
    public void onServiceDisconnected(ComponentName name) {
            }
};

@Override
protected void onCreate(Bundle savedInstanceState) {
    super.onCreate(savedInstanceState);
    setContentView(R.layout.activity_main);
    //获取界面中的按钮
    Button start = (Button) findViewById(R.id.btn_start);
    Button stop = (Button) findViewById(R.id.btn_stop);
    Button bindStart = (Button) findViewById(R.id.btn_bind_start);
    Button bindStop = (Button) findViewById(R.id.btn_bind_stop);
    Button showText = (Button) findViewById(R.id.btn_show);
    mtvCount = (TextView) findViewById(R.id.tv_count);
    //创建启动 Service 的 Intent
    final Intent intent = new Intent(this, MyBindService.class);
    start.setOnClickListener(new View.OnClickListener() {
        @Override
        public void onClick(View view) {
            //启动 Service
            startService(intent);
        }
    });
    stop.setOnClickListener(new View.OnClickListener() {
        @Override
        public void onClick(View view) {
            //关闭 Service
            stopService(intent);
```

```
                }
            });
            bindStart.setOnClickListener(new View.OnClickListener() {
                @Override
                public void onClick(View view) {
                    // 绑定指定的 Service
                    bindService(intent, conn, Service.BIND_AUTO_CREATE);
                }
            });
            bindStop.setOnClickListener(new View.OnClickListener() {
                @Override
                public void onClick(View view) {

                    // 解除绑定 Service
                    unbindService(conn);
                }
            });
            showText.setOnClickListener(new View.OnClickListener() {
                @Override
                public void onClick(View source) {
                    mtvCount.setText("Service 的 count 值: " + binder.getCount());
                }
            });
        }
    }
```

在 bindService()方法中接收三个参数，第一个参数就是刚刚构建出的 Intent 对象。第二个参数是前面创建的 ServiceConnection 的实例，该对象主要用于监听访问者与 Service 之间的连接情况。当访问者与 Service 之间连接成功时，将回调该 ServiceConnection 对象的 onServiceConnected()方法；当 Service 所在的宿主进程由于异常中止或其他原因中止，导致 Service 与访问者之间断开连接时回调 ServiceConnection 对象的 onServiceDisconnected()方法。第三个参数是一个标志位，指绑定的时候是否自动创建 Service，前提是没有创建 Service。这里传入 BIND_AUTO_CREATE 表示在 Activity 和 Service 建立关联后自动创建 Service，这会使得 MyService 中的 onCreate()方法得到执行，但 onStartCommand()方法不会执行。

这里的 ServiceConnection 对象的 onServiceDisconnected()方法中有一个 IBinder 对象，该对象就可以实现与被绑定的 Service 之间的通信。

运行一下程序，单击“启动 BindService”按钮，再单击“显示数据”按钮就可以看到 Service 的运行状态。单击“关闭 BindService”就可以关闭 Service。

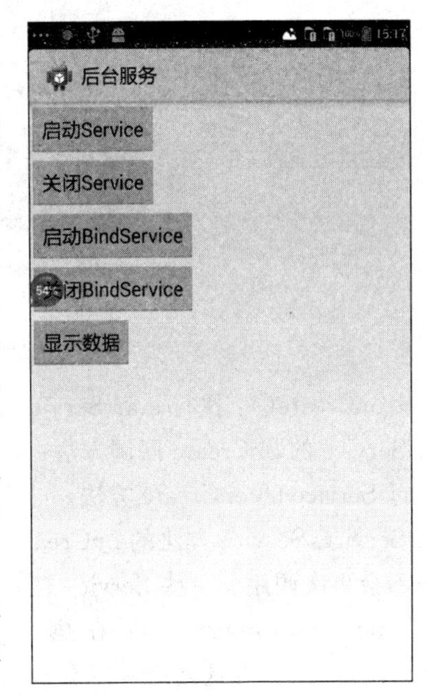

图 6-3 界面中显示的各按钮

效果如图 6-3、图 6-4 所示。

com.qsd.ch6_2	MyBindService	Created()
com.qsd.ch6_2	MyBindService	onBind()
com.qsd.ch6_2	MyBindService	onUnbind()
com.qsd.ch6_2	MyBindService	onDestroy()

图 6-4 Service 的运行状态

6.3 Service 的生命周期

通过前面的例子我们大致了解了 Service 的基本用法。随着应用程序启动 Service 方式的不同，Service 的生命周期也会略有不同，如图 6-5 所示。

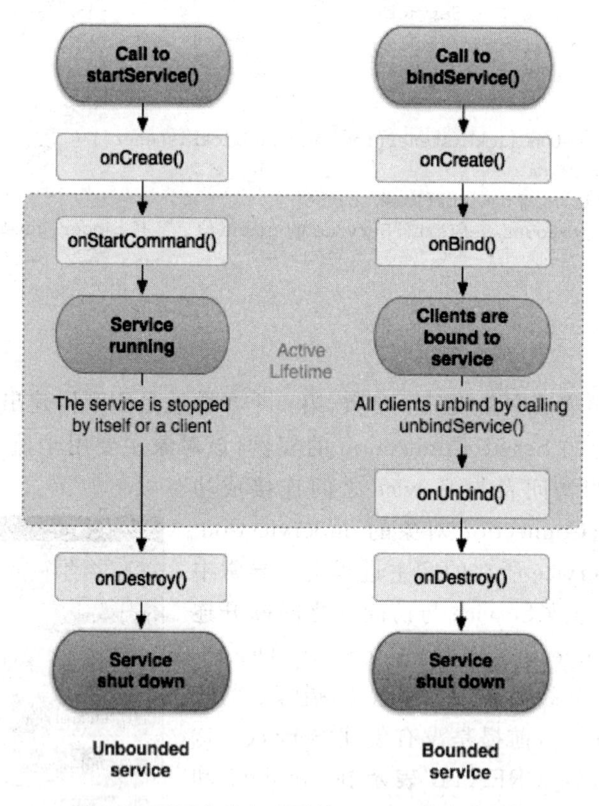

图 6-5 不同启动方式下的 Service 的生命周期

onCreate()：执行 startService 方法时，如果 Service 没有运行则会创建该 Service，并执行 Service 的 onCreate 回调方法；如果 Service 已经处于运行中，那么执行 startService 不会执行 Service 的 onCreate 方法。也就是说如果多次执行了 Context 的 startService 方法启动 Service，Service 方法的 onCreate 方法只会在第一次创建 Service 的时候调用一次，以后均不会再次调用。一些 Service 初始化的相关操作可以在 onCreate 方法中完成。

onStartCommand( )：在执行了 startService 方法之后，有可能会调用 Service 的 onCreate 方法，在这之后一定会执行 Service 的 onStartCommand 回调方法。也就是说，如果多次执行了 Context 的 startService 方法，那么 Service 的 onStartCommand 方法也会相

应被多次调用。onStartCommand()方法很重要，在该方法中可根据传入的 Intent 参数进行实际的操作，例如在此处创建一个线程用于下载数据或播放音乐等。

onBindService()：当其他组件需要通过 bindService()绑定服务时，例如执行远程进程调用 RPC(Remote Procedure Call)，系统会调用 onBindService()方法。在本方法的实现代码中，必须返回 IBinder 来提供一个接口，客户端使用它和服务进行通信。必须确保实现本方法。不过如不需要提供绑定，返回 null 即可。

onDestroy()：当服务不用了并要被销毁时，系统会调用 onDestroy()方法。这是服务收到的最后一个调用。应该实现本方法来进行资源(诸如线程、已注册的监听器 listener 和接收器 receiver 等)的清理工作。

6.4 Service 的其他用法

6.4.1 使用前台服务

服务是一个应用程序组件，它代表应用程序希望较长时间的运行、操作而不与用户交互，或提供功能供其他应用程序使用。

每个服务类必须有一个相应的节点注册在清单文件 AndroidManifest. xml 中，用 < Service >声明。

服务可以通过 Context. startService()和 Context. bindService()开始工作，服务和其他的应用对象一样，在它的宿主进程的主线程中运行。

注意：与其他应用程序对象一样，服务在其宿主进程的主线程中运行。这意味着，如果你的服务要做任何 CPU 密集型(如 MP3 播放)或阻塞(如网络)的操作，它会以它自己的线程去做这项工作。更多的信息可以在进程和线程中找到。

有些项目因为特殊的需求会要求必须使用前台服务，例如天气预报的 Service 在后台更新数据的同时，还会在系统状态栏一直显示当前的信息，如图 6-6 所示。

图 6-6 天气预报的 Service 会一直显示当前的信息

为使用前台服务，添加一个 send()方法，在 send()方法中添加一个通知栏对象。主要代码如下：

```
package com.qsd.ch6_4_1;

import android.annotation.SuppressLint;
import android.app.Activity;
import android.app.Notification;
import android.app.NotificationManager;
import android.app.PendingIntent;
import android.app.Service;
import android.content.ComponentName;
```

```
import android.content.Intent;
import android.content.ServiceConnection;
import android.os.Bundle;
import android.os.IBinder;
import android.view.View;
import android.widget.Button;
import android.widget.TextView;

import com.qsd.ch6_3.R;

@SuppressLint("NewApi")
public class MainActivity extends Activity {
    NotificationManager nm;
    // 保持所启动的 Service 的 IBinder 对象
    TextView mtvCount;
    MyBindService.MyBinder binder;
    // 定义一个 ServiceConnection 对象
    private ServiceConnection conn = new ServiceConnection() {
        // 当该 Activity 与 Service 连接成功时回调该方法
        @Override
        public void onServiceConnected(ComponentName name, IBinder service) {
            System.out.println("--Service Connected--");
            // 获取 Service 的 onBind 方法所返回的 MyBinder 对象
            binder = (MyBindService.MyBinder) service;
            send();
        }

        // 当该 Activity 与 Service 断开连接时回调该方法
        @Override
        public void onServiceDisconnected(ComponentName name) {
            System.out.println("--Service Disconnected--");
        }
    };

    @Override
    protected void onCreate(Bundle savedInstanceState) {
        super.onCreate(savedInstanceState);
        setContentView(R.layout.activity_main_bind);
        // 获取界面中的按钮
        Button start = (Button) findViewById(R.id.btn_start);
        Button stop = (Button) findViewById(R.id.btn_stop);
        Button bindStart = (Button) findViewById(R.id.btn_bind_start);
        Button bindStop = (Button) findViewById(R.id.btn_bind_stop);
        Button showText = (Button) findViewById(R.id.btn_show);
        mtvCount = (TextView) findViewById(R.id.tv_count);
        // 创建启动 Service 的 Intent
        final Intent intent = new Intent(this, MyBindService.class);
        start.setOnClickListener(new View.OnClickListener() {
            @Override
            public void onClick(View view) {
                // 启动 Service
```

```
                startService(intent);
            }
        });
        stop.setOnClickListener(new View.OnClickListener() {
            @Override
            public void onClick(View view) {
                // 关闭 Service
                stopService(intent);
            }
        });
        bindStart.setOnClickListener(new View.OnClickListener() {
            @Override
            public void onClick(View view) {
                // 绑定指定 Service
                bindService(intent, conn, Service.BIND_AUTO_CREATE);
            }
        });
        bindStop.setOnClickListener(new View.OnClickListener() {
            @Override
            public void onClick(View view) {

                // 解除绑定 Service
                unbindService(conn);
            }
        });
        showText.setOnClickListener(new View.OnClickListener() {
            @Override
            public void onClick(View source) {
                mtvCount.setText("Service 的 count 的值: " + binder.getCount());
            }
        });
        nm = (NotificationManager) getSystemService(NOTIFICATION_SERVICE);
        showText.setOnClickListener(new View.OnClickListener() {
            @Override
            public void onClick(View view) {
                send();
            }
        });
    }

    /***
     * 发送通知
     */
    private void send() {
        // 创建一个启动其他 Activity 的 Intent
        Intent intent = new Intent(MainActivity.this, FirstActivity.class);
        PendingIntent pi = PendingIntent.getActivity(MainActivity.this, 0,
                intent, 0);
        Notification notify = new Notification.Builder(MainActivity.this)
        // 设置打开该通知,该通知自动消失
                .setAutoCancel(true)
                // 设置显示在状态栏的通知提示信息
```

```
                .setTicker("有新消息")
                // 设置通知小图标
                .setSmallIcon(R.drawable.ic_launcher)
                // 设置通知标题内容
                .setContentTitle("一条新通知")
                // 设置通知内容
                .setContentText("恭喜你,你会使用前台服务了")
                // 设置通知的自定义时间

                .setWhen(System.currentTimeMillis())
                // 设置通知将要启动程序的 Intent
                .setContentIntent(pi).build();
        // 发送通知
        nm.notify(0x123, notify);
    }

}
```

运行之后,单击“显示数据”,效果如图 6-7 所示。

6.4.2 使用 IntentService

IntentService 是 Service 类的子类,它使用工作(worker)线程来处理所有的启动请求,每次请求都会启动一个线程。如果服务不需要同时处理多个请求,这是最佳的选择。所有要做的工作就是实现 onHandleIntent()即可,它会接收每个启动请求的 Intent,然后在后台完成工作。

图 6-7 显示效果

IntentService 将执行以下步骤:

(1) 创建一个默认的工作(worker)线程,它独立于应用程序主线程来执行所有发送到 onStartCommand()的 Intent。

(2) 创建一个工作队列,每次向 onHandleIntent()传入一个 Intent,这样就不必担心多线程问题了。

(3) 在处理完所有的启动请求后,终止服务,这样就不需调用 stopSelf()了;提供默认的 onBind()实现代码,它返回 null;提供默认的 onStartCommand()实现代码,把 Intent 送入工作队列,稍后会再传给 onHandleIntent()实现代码。

以上所有步骤将汇成一个结果:要做的全部工作就是实现 onHandleIntent()的代码,来完成客户端提交的任务(当然还需要为服务提供一小段构造方法)。以下是一个 IntentService 的实现例程:

```
public class MyIntentService extends IntentService {

public MyIntentService() {
    super("MyIntentService");
}
```

```
/**
  * IntentService从默认的工作线程中调用本方法,并用启动服务的intent作为参数.
  * 本方法返回后,IntentService将适时终止这个服务.
  */
  @Override
  protected void onHandleIntent(Intent intent) {
// 通常会在这里执行一些工作,例如下载文件等

   Log.e("MyIntentService", "线程的标识符:" + Thread.currentThread().getId());
  }

  @Override
  public void onDestroy() {
      super.onDestroy();

      Log.e("MyIntentService", "onDestroy()");
  }
}
```

代码中构造方法是必需的,必须用工作线程名称作为参数。我们实现了onDestroy()方法,在里面添加一行代码Log用于测试。修改MainActivity程序中的代码,下面列出其主要代码。

activity_intent_service.xml代码:

```
  <?xml version="1.0" encoding="utf-8"?>
<LinearLayout xmlns:android="http://schemas.android.com/apk/res/android"
  xmlns:tools="http://schemas.android.com/tools"
  android:layout_width="match_parent"
  android:layout_height="match_parent"
  android:orientation="vertical"
  >

  <Button
      android:id="@+id/btn_start"
      android:layout_width="wrap_content"
      android:layout_height="wrap_content"
      android:text="启动Service" />

  <Button
      android:id="@+id/btn_stop"
      android:layout_width="wrap_content"
      android:layout_height="wrap_content"
      android:text="关闭Service" />

  <Button
      android:id="@+id/btn_bind_start"
      android:layout_width="wrap_content"
      android:layout_height="wrap_content"
      android:text="启动BindService" />
```

```
    <Button
        android:id="@+id/btn_bind_stop"
        android:layout_width="wrap_content"
        android:layout_height="wrap_content"
        android:text="关闭 BindService" />

    <Button
        android:id="@+id/btn_intnet_service"
        android:layout_width="wrap_content"
        android:layout_height="wrap_content"
        android:text="启动 IntentService" />

    <Button
        android:id="@+id/btn_show"
        android:layout_width="wrap_content"
        android:layout_height="wrap_content"
        android:text="显示数据" />

    <TextView
        android:id="@+id/tv_count"
        android:layout_width="wrap_content"
        android:layout_height="wrap_content"
        android:layout_gravity="center_horizontal"
        android:padding="20dp"
        android:textColor="#050f98" />

</LinearLayout>
```

添加一个 id 是 btn_intnet_service 的按钮。MainActivity 代码如下：

```
package com.qsd.ch6_4_2;

import android.annotation.SuppressLint;
import android.app.Activity;
import android.app.Notification;
import android.app.NotificationManager;
import android.app.PendingIntent;
import android.app.Service;
import android.content.ComponentName;
import android.content.Intent;
import android.content.ServiceConnection;
import android.os.Bundle;
import android.os.IBinder;
import android.util.Log;
import android.view.View;
import android.widget.Button;
import android.widget.TextView;

import com.qsd.ch6_4.R;
import com.qsd.ch6_4_1.MyBindService;
```

```
@SuppressLint("NewApi")
public class MainActivity extends Activity {
    NotificationManager nm;
    // 保持所启动的 Service 的 IBinder 对象
    TextView mtvCount;
    MyBindService.MyBinder binder;
    // 定义一个 ServiceConnection 对象
    private ServiceConnection conn = new ServiceConnection() {
        // 当该 Activity 与 Service 连接成功时回调该方法
        @Override
        public void onServiceConnected(ComponentName name, IBinder service) {
            System.out.println(" -- Service Connected -- ");
            // 获取 Service 的 onBind 方法所返回的 MyBinder 对象
            binder = (MyBindService.MyBinder) service;
            send();
        }

        // 当该 Activity 与 Service 断开连接时回调该方法
        @Override
        public void onServiceDisconnected(ComponentName name) {
            System.out.println(" -- Service Disconnected -- ");
        }
    };

    @Override
    protected void onCreate(Bundle savedInstanceState) {
        super.onCreate(savedInstanceState);
        setContentView(R.layout.activity_main);
        nm = (NotificationManager) getSystemService(NOTIFICATION_SERVICE);
        // 获取界面中的按钮
        Button start = (Button) findViewById(R.id.btn_start);
        Button stop = (Button) findViewById(R.id.btn_stop);
        Button bindStart = (Button) findViewById(R.id.btn_bind_start);
        Button bindStop = (Button) findViewById(R.id.btn_bind_stop);
        Button intentService = (Button) findViewById(R.id.btn_intnet_service);
        mtvCount = (TextView) findViewById(R.id.tv_count);
        // 创建启动 Service 的 Intent
        final Intent intent = new Intent(this, MyBindService.class);

        start.setOnClickListener(new View.OnClickListener() {
            @Override
            public void onClick(View view) {
                // 启动 Service
                startService(intent);
            }
        });
        stop.setOnClickListener(new View.OnClickListener() {
            @Override
            public void onClick(View view) {
                // 关闭 Service
                stopService(intent);
```

```
            }
        });
        bindStart.setOnClickListener(new View.OnClickListener() {
            @Override
            public void onClick(View view) {
                // 绑定指定的 Service
                bindService(intent, conn, Service.BIND_AUTO_CREATE);
            }
        });
        bindStop.setOnClickListener(new View.OnClickListener() {
            @Override
            public void onClick(View view) {
                // 解除绑定 Service
                unbindService(conn);
            }
        });
        intentService.setOnClickListener(new View.OnClickListener() {
            @Override
            public void onClick(View view) {
                Log.e("MainActivity", "线程的标识符:"
                        + Thread.currentThread().getId());
                Intent intentService = new Intent(MainActivity.this,
                        MyIntentService.class);
                startService(intentService);
            }
        });
        mtvCount.setOnClickListener(new View.OnClickListener() {
            @Override
            public void onClick(View view) {
                send();
            }
        });
    }

    /***
     * 发送通知
     */
    private void send() {
        // 创建一个启动其他 Activity 的 Intent
        Intent intent = new Intent(MainActivity.this, FirstActivity.class);
        PendingIntent pi = PendingIntent.getActivity(MainActivity.this, 0,
                intent, 0);
        Notification notify = new Notification.Builder(MainActivity.this)
                // 设置打开该通知,该通知自动消失
                .setAutoCancel(true)
                // 设置显示在状态栏的通知提示信息
                .setTicker("有新消息")
                // 设置通知的小图标
                .setSmallIcon(R.drawable.ic_launcher)
                // 设置通知内容的标题
                .setContentTitle("一条新通知")
                // 设置通知内容
                .setContentText("恭喜你,你会使用前台服务了")
```

```
                // 设置使用系统默认的声音和默认的 LED 灯
                // .setDefaults(Notification.DEFAULT_SOUND
                // |Notification.DEFAULT_LIGHTS)
                // 设置通知的自定义声音
                .setWhen(System.currentTimeMillis())
                // 设置通知将要启动程序的 Intent
                .setContentIntent(pi) //
                .build();
        // 发送通知
        nm.notify(0x123, notify);
    }

}
```

在这里给新添加的按钮添加一个事件并启动 MyIntentService 服务,这样可以打印一个主线程的 id,用于与之后的 IntentService 进行比较。重新运行之后的效果如图 6-8 所示。

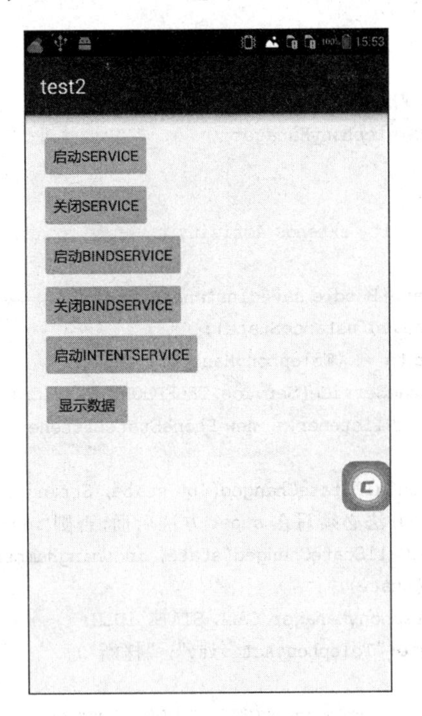

图 6-8 重新运行之后的效果

单击“启动 INTENTSERVICE”按钮,观察 LogCat 中打印的日志,如图 6-9 所示。

com.qsd.ch6_4	MainActivity	线程的标识符:1
com.qsd.ch6_4	MyIntentService	线程的标识符:2776
com.qsd.ch6_4	MyIntentService	onDestroy()

图 6-9 LogCat 中的打印日志

从效果图中可以看到,MyIntentService 和 MainActivity 所在的线程 id 不一样,而且 onDestroy()方法也得到了执行,说明 MyIntentService 在运行完毕之后确实是自动停止了,比之前手动关闭好了不少。

6.5 常见的系统服务

前面用到的 Service 都是在 APP 里实现的。如何调用系统的服务获取其中的内容，减少用户输入次数呢？下面就开始学习系统服务。

6.5.1 电话管理器

电话管理器 TelephonyManager 是一个管理手机通话状态、电话网络信息的服务类，该类提供了大量的 getXXX ()方法来获取电话网络的相关信息。

新建一个 TelephonyActivity 类，主要代码如下：

```
package com.qsd;
import android.app.Activity;
import android.app.Service;
import android.os.Bundle;
import android.telephony.PhoneStateListener;
import android.telephony.TelephonyManager;
import android.util.Log;

public class TelephonyActivity extends Activity {
    @Override
    protected void onCreate(Bundle savedInstanceState) {
        super.onCreate(savedInstanceState);
        TelephonyManager tm = (TelephonyManager) this
                .getSystemService(Service.TELEPHONY_SERVICE);
        PhoneStateListener listener = new PhoneStateListener() {
            @Override
            public void onCallStateChanged(int state, String incomingNumber) {
                // 注意,方法必须写在 super 方法后面,否则 incomingNumber 无法获取值
                super.onCallStateChanged(state, incomingNumber);
                switch (state) {
                case TelephonyManager.CALL_STATE_IDLE:
                    Log.e("TelephonyActivity", "挂断");
                    break;
                case TelephonyManager.CALL_STATE_OFFHOOK:
                    Log.e("TelephonyActivity", "接听");
                    break;
                case TelephonyManager.CALL_STATE_RINGING:// 输出来电号码
                    Log.e("TelephonyActivity", "响铃:来电号码" + incomingNumber);
                    break;
                }
            }
        };
        // 设置一个监听器
        tm.listen(listener, PhoneStateListener.LISTEN_CALL_STATE);
    }
}
```

想要监听电话的拨打状况，需要以下几步：

(1) 获取电话服务管理器 TelephonyManager：

```
manager = this.getSystemService(TELEPHONY_SERVICE);
```

(2) 通过 TelephonyManager 注册要监听的电话状态改变事件：

```
manager.listen(new MyPhoneStateListener(),PhoneStateListener
  .LISTEN_CALL_STATE);
```

其中的 PhoneStateListener.LISTEN_CALL_STATE 就是要监听的状态改变事件。

(3) 通过 extends PhoneStateListener 定制自己的规则，将其对象传递给第(2)步作为参数。

(4) 给应用添加权限：

```
android.permission.READ_PHONE_STATE…
```

6.5.2 短信管理器

短信管理器 SmsManager 是 Android 提供的另一个常见的服务，SmsManager 提供了一系列 sendXXXMessage()方法用于发送短信。通常发送短信都是调用 sendTextMessage()方法发送。

新建 SmsManagerActivity，主要代码如下：

```
package com.qsd;

import java.util.ArrayList;

import android.app.Activity;
import android.app.PendingIntent;
import android.content.Intent;
import android.os.Bundle;
import android.telephony.SmsManager;
import android.view.View;
import android.widget.Button;
import android.widget.EditText;
import android.widget.Toast;

import com.qsd.ch6_5.R;

public class SmsManagerActivity extends Activity {
    EditText mphone, mcontext;
    SmsManager smsManager;

    @Override
    protected void onCreate(Bundle savedInstanceState) {
        super.onCreate(savedInstanceState);
        setContentView(R.layout.activity_sms_manager);
        // 获取 SmsManager 对象
        smsManager = SmsManager.getDefault();
```

```
        mphone = (EditText) findViewById(R.id.edt_phone);
        mcontext = (EditText) findViewById(R.id.edt_context);
        Button btn_send = (Button) findViewById(R.id.btn_send);

        btn_send.setOnClickListener(new View.OnClickListener() {
            @Override
            public void onClick(View view) {
                String phone = mphone.getText().toString();
                String context = mcontext.getText().toString();
                if (phone.isEmpty()) {
                Toast.makeText(SmsManagerActivity.this, "请填写手机号",
                            Toast.LENGTH_LONG).show();
                    return;
                }
                ArrayList<String> list = smsManager.divideMessage(context);
                // 因为一条短信有字数限制,因此要将长短信拆分
                PendingIntent pi = PendingIntent.getActivity(
                        SmsManagerActivity.this, 0, new Intent(), 0);
                for (String text : list) {
                    smsManager.sendTextMessage(phone, null, text, pi, null);
                }
                Toast.makeText(SmsManagerActivity.this, "发送成功",
                        Toast.LENGTH_LONG).show();
            }
        });
    }
}
```

activity_sms_manager.xml 代码如下所示:

```
<?xml version="1.0" encoding="utf-8"?>
<LinearLayout xmlns:android="http://schemas.android.com/apk/res/android"
    xmlns:tools="http://schemas.android.com/tools"
    android:layout_width="match_parent"
    android:layout_height="match_parent"
    android:orientation="vertical">
    <TextView
        android:layout_width="wrap_content"
        android:layout_height="wrap_content"
        android:layout_margin="10dp"
        android:text="请输入手机号"
        android:textSize="24sp" />

    <EditText
        android:id="@+id/edt_phone"
        android:layout_width="match_parent"
        android:layout_height="wrap_content" />

    <TextView
        android:layout_width="wrap_content"
        android:layout_height="wrap_content"
```

```
        android:layout_margin = "10dp"
        android:text = "请输入短信内容"
        android:textSize = "24sp" />

    <EditText
        android:id = "@ + id/edt_context"
        android:layout_width = "match_parent"
        android:layout_height = "wrap_content" />

    <Button
        android:id = "@ + id/btn_send"
        android:layout_width = "wrap_content"
        android:layout_height = "wrap_content"
        android:text = "发送" />
</LinearLayout>
```

上边代码中写了两个文本框、两个输入框和一个按钮。在SmsManagerActivity代码中用到了一个PendingIntent对象。PendingIntent对象是对Intent的包装，一般通过调用PendingIntent的getActivity()、getService()、getBrodcastReceiver()静态方法来获取PendingIntent对象。

与Intent不同的是，PendingIntent对象通常会传给其他应用组件，从而由其他应用程序来执行PendingIntent包装的"Intent"。

因为该程序需要调用SmsManager发送短信，所以需要获取该程序发送短信的权限。之前是在AndroidManifest.xml中添加权限，现在用代码动态添加：

```
ActivityCompat.requestPermissions(this,new String[]{Manifest
  .permission.SEND_SMS},1)
```

因为此权限在API23之后就不能静态添加了。

6.5.3 振动器

振动器Vibrator其实就是Android提供的用于机身震动的一个服务。如果程序退出，程序引发的任何振动都会停止，例如QQ的震动提醒。

Vibrator三个常用的方法：

- void android.os.Vibrator.vibrate(long milliseconds)：震动milliseconds毫秒。
- void android.os.Vibrator.vibrate(long[] pattern, int repeat)：传递一个int数组，它们是以毫秒为单位打开或关闭振动器的持续时间。第一个值表示在打开振动器之前要等待的毫秒数，第二个值表示在关闭振动器之前保持振动器的毫秒数，随后的值交替执行，以关闭振动器或打开振动器。接着就从pattern[int repeat]的位置开始重复，一直重复下去。传入的repeat为-1是不重复震动，传入repeat为0是一直重复震动(下标为0的数值是等待时间，下标为1的数值是震动时间)，传入repeat为1是从数组long pattern[]中下标为1的地方开始震动(下标为1的数值是等待时间，下标为2的数值是震动时间)，传入repeat为2是从数组long pattern[]中下标为2的地方开始震动(下标为2的数值是等待时间，下标为3的数值是震动时间)，以此类推。

- void android. os. Vibrator. cancel()：关闭手机震动。

注意：在使用 Vibrator 类时，要在清单文件中声明 VIBRATE 权限：

```
<uses-permission android:name="android.permission.VIBRATE" />.
```

下面是一个简单的振动器。新建一个 VibratorActivity 类，主要代码如下所示：

```
package com.qsd;

import android.app.Activity;
import android.app.Service;
import android.os.Bundle;
import android.os.Vibrator;
import android.view.View;
import android.widget.Button;

import com.qsd.ch6_5.R;

public class VibratorActivity extends Activity {

    @Override
    protected void onCreate(Bundle savedInstanceState) {
        super.onCreate(savedInstanceState);
        setContentView(R.layout.activity_vibrator);
        Button mstart = (Button) findViewById(R.id.btn_start);
        Button mend = (Button) findViewById(R.id.btn_stop);
        final Vibrator mvibrator
= (Vibrator) getSystemService(Service.VIBRATOR_SERVICE);
        mstart.setOnClickListener(new View.OnClickListener() {
            @Override
            public void onClick(View view) {
                // 控制手机震动 2 秒
                mvibrator.vibrate(new long[] { 400, 800, 1200, 1600 }, 2);
            }
        });
        mend.setOnClickListener(new View.OnClickListener() {
            @Override
            public void onClick(View view) {
                // 关闭振动
                mvibrator.cancel();
            }
        });
    }
}
```

xml 主要代码如下所示：

```
<?xml version="1.0" encoding="utf-8"?>
<LinearLayout xmlns:android="http://schemas.android.com/apk/res/android"
    xmlns:tools="http://schemas.android.com/tools"
```

```
    android:layout_width = "match_parent"
    android:layout_height = "match_parent"
    android:orientation = "horizontal"
    >

    <Button
        android:id = "@ + id/btn_start"
        android:layout_width = "wrap_content"
        android:layout_height = "wrap_content"
        android:text = "启动" />

    <Button
        android:id = "@ + id/btn_stop"
        android:layout_width = "wrap_content"
        android:layout_height = "wrap_content"
        android:text = "关闭" />
</LinearLayout>
```

代码很简单,就是用“启动”和“关闭”这两个按钮来控制振动。由于程序控制手机振动需要得到相应的权限,因此要在清单文件 AndroidManifest. xml 中添加如下代码:

```
<uses - permission android:name = "android.permission.VIBRATE" />
```

由于模拟器运行无法查看振动效果,所以要在真机上运行测试。

6.5.4 闹钟/全局定时器

闹钟/全局定时器 AlarmManager 是 Android 中常用的一种系统级别的警告服务,也称作全局定时器。这些服务允许安排应用程序在将来某个时候运行。它可以用来开发手机闹铃,也可以在指定时间或者指定周期内启动其他如 Activity、Service、BroadcastReceiver 之类的组件。

AlarmManager 的常用方法有三个:

(1) set(int type,long startTime,PendingIntent pi):该方法用于设置一次性闹钟。第一个参数表示闹钟类型,第二个参数表示闹钟执行时间,第三个参数表示闹钟响应动作。

(2) setRepeating(int type,long startTime,long intervalTime,PendingIntent pi):该方法用于设置重复闹钟。第一个参数表示闹钟类型,第二个参数表示闹钟首次执行时间,第三个参数表示闹钟两次执行的间隔时间,第四个参数表示闹钟响应的动作。

(3) setInexactRepeating(int type,long startTime,long intervalTime,PendingIntent pi):该方法也用于设置重复闹钟。与第二个方法相似,不过其闹钟两次执行的间隔时间不是固定的。

三个方法各参数的说明如下:

(1) int type:闹钟的类型。常用的有 AlarmManager. ELAPSED_REALTIME、AlarmManager. ELAPSED_REALTIME_WAKEU、AlarmManager. RTC、AlarmManager. RTC_WAKEUP、AlarmManager. POWER_OFF_WAKEUP 这 5 个值。其中,

AlarmManager. ELAPSED_REALTIME 表示闹钟在手机睡眠状态下不可用。该状态

下闹钟使用相对时间(相对于系统启动开始),状态值为3;

AlarmManager.ELAPSED_REALTIME_WAKEUP表示闹钟在睡眠状态下会唤醒系统并执行提示功能。该状态下闹钟也使用相对时间,状态值为2;

AlarmManager.RTC表示闹钟在睡眠状态下不可用。该状态下闹钟使用绝对时间,即当前系统时间,状态值为1;

AlarmManager.RTC_WAKEUP表示闹钟在睡眠状态下会唤醒系统并执行提示功能。该状态下闹钟使用绝对时间,状态值为0;

AlarmManager.POWER_OFF_WAKEUP表示闹钟在手机关机状态下也能正常进行提示,所以是5个状态中使用最多的状态之一,该状态下闹钟也是用绝对时间,状态值为4。不过这个状态好像受SDK版本影响,某些版本并不支持。

(2) long startTime:闹钟的第一次执行时间,以毫秒为单位。可以自定义时间,不过一般使用当前时间。需要注意的是,本参数与第一个参数(type)密切相关。如果第一个参数对应的闹钟使用的是相对时间(ELAPSED_REALTIME和ELAPSED_REALTIME_WAKEUP),那么本参数就得使用相对时间(相对于系统启动时间来说),如当前时间就表示为SystemClock.elapsedRealtime();如果第一个参数对应的闹钟使用的是绝对时间(RTC、RTC_WAKEUP、POWER_OFF_WAKEUP),那么本参数就得使用绝对时间,如当前时间就表示为System.currentTimeMillis()。

(3) long intervalTime:只有后两个方法存在本参数,表示两次闹钟执行的间隔时间,也是以毫秒为单位。

(4) PendingIntent pi:绑定了闹钟的执行动作,例如发送一个广播、给出提示等。

下面是一个简单的例子。新建一个AlarmManagerActivity,代码如下:

```
package com.qsd;

import java.util.Calendar;

import android.app.Activity;
import android.app.AlarmManager;
import android.app.PendingIntent;
import android.app.TimePickerDialog;
import android.content.Intent;
import android.os.Bundle;
import android.util.Log;
import android.view.View;
import android.widget.Button;
import android.widget.TimePicker;
import android.widget.Toast;

import com.qsd.ch6_5.R;

public class AlarmManagerActivity extends Activity {
    private Button btnSetClock;
    private Button btnbtnCloseClock;
    private AlarmManager alarmManager;
    private PendingIntent pi;
```

```
    @Override
    protected void onCreate(Bundle savedInstanceState) {
        super.onCreate(savedInstanceState);
        setContentView(R.layout.activity_alarm_manager);
        startAlarm();
    }

    private void startAlarm() {
        btnSetClock = (Button) findViewById(R.id.btnSetClock);
        btnbtnCloseClock = (Button) findViewById(R.id.btnCloseClock);

        // 获取AlarmManager对象
        alarmManager = (AlarmManager) getSystemService(ALARM_SERVICE);
        // 指定要启动的是Activity组件,通过PendingIntent调用getActivity来设置
        Intent intent = new Intent(AlarmManagerActivity.this,
                ClockActivity.class);
        pi = PendingIntent.getActivity(AlarmManagerActivity.this, 0, intent, 0);
        btnSetClock.setOnClickListener(new View.OnClickListener() {
            @Override
            public void onClick(View v) {
                Calendar currentTime = Calendar.getInstance();
                // 弹出一个设置时间的对话框,供用户选择时间
                new TimePickerDialog(AlarmManagerActivity.this, 0,
                        new TimePickerDialog.OnTimeSetListener() {
                            @Override
                            public void onTimeSet(TimePicker view,
                                    int hourOfDay, int minute) {
                                // 设置当前时间
                                Calendar c = Calendar.getInstance();
                                c.setTimeInMillis(System.currentTimeMillis());
                                // 根据用户选择的时间来设置Calendar对象
                                c.set(Calendar.HOUR_OF_DAY, hourOfDay);
                                c.set(Calendar.MINUTE, minute);
                        // 设置AlarmManager在Calendar对应的时间启动Activity
// AlarmManager.RTC_WAKEUP设置这个参数意味着即使系统处于关机状态,
//到了系统预定时间,AlarmManager也会控制系统去执行pi对应的Activity组件
                                alarmManager.set(AlarmManager.RTC_WAKEUP,
                                        c.getTimeInMillis(), pi);
                                Log.e("shijian:", c.getTimeInMillis() + "");
                                // 提示闹钟设置完毕
                                Toast.makeText(AlarmManagerActivity.this,
                                " 闹钟设置完毕", Toast.LENGTH_SHORT).show();
                            }
                        }, currentTime.get(Calendar.HOUR_OF_DAY), currentTime
                                .get(Calendar.MINUTE), false).show();
                btnbtnCloseClock.setVisibility(View.VISIBLE);
            }
        });

        btnbtnCloseClock.setOnClickListener(new View.OnClickListener() {
```

```
                @Override
                public void onClick(View v) {
                    alarmManager.cancel(pi);
                    btnbtnCloseClock.setVisibility(View.GONE);
                    Toast.makeText(AlarmManagerActivity.this, "闹钟已取消",
                            Toast.LENGTH_SHORT).show();
                }
            });
        }
    }
```

activity_alarm_manager.xml 主要代码如下所示：

```
<?xml version="1.0" encoding="utf-8"?>
<LinearLayout xmlns:android="http://schemas.android.com/apk/res/android"
    xmlns:tools="http://schemas.android.com/tools"
    android:layout_width="match_parent"
    android:layout_height="match_parent"
    android:orientation="horizontal"
    >

    <Button
        android:id="@+id/btnSetClock"
        android:layout_width="wrap_content"
        android:layout_height="wrap_content"
        android:text="设置闹铃" />

    <Button
        android:id="@+id/btnCloseClock"
        android:layout_width="wrap_content"
        android:layout_height="wrap_content"
        android:text="关闭闹铃" />

</LinearLayout>
```

上面的程序启动了一个名为ClockActivity的Activity，它非常简单，不需要程序界面。当该Activity加载时会打开一个对话框提示闹铃时间到并播放一段音乐提醒用户。

主要代码如下所示：

```
package com.qsd;

import android.app.Activity;
import android.app.AlertDialog;
import android.content.DialogInterface;
import android.media.MediaPlayer;
import android.os.Bundle;

import com.qsd.ch6_5.R;

public class ClockActivity extends Activity {
```

```
    private MediaPlayer mediaPlayer;

    @Override
    protected void onCreate(Bundle savedInstanceState) {
        super.onCreate(savedInstanceState);

        mediaPlayer = MediaPlayer.create(this, R.raw.alarm);
        // mediaPlayer.setLooping(true);
        mediaPlayer.start();

        // 创建一个闹钟提醒的对话框,单击"关闭闹铃"关闭铃声与页面
        new AlertDialog.Builder(ClockActivity.this)
                .setTitle("闹钟")
                .setMessage("起床洗脸刷牙了")
                .setPositiveButton("关闭闹铃",
                        new DialogInterface.OnClickListener() {
                            @Override
                            public void onClick(DialogInterface dialog,
                                    int which) {
                                mediaPlayer.stop();
                                ClockActivity.this.finish();
                            }
                        }).show();
    }
}
```

运行程序,单击"设置闹铃"按钮,之后在弹出的对话框中单击"确定"按钮。运行效果如图 6-10 所示。

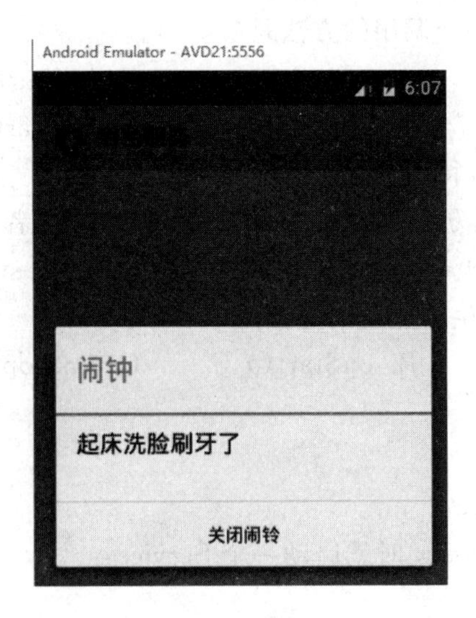

图 6-10 闹钟运行效果

注意:要在清单文件 AndroidManifest.xml 中配置 Activity,否则不会显示 ClockActivity 这个活动。

6.6 本章小结

本章内容主要介绍四大组件之一的 Service。介绍了 Service 的创建和配置以及启动和关闭，Service 与 Activity 的通信。介绍了前台服务的使用和系统服务的使用。本章重点是 Service 的启动与配置方法以及系统服务的调用。

6.7 练习题

一、填空题

1. 创建服务的时候必须要继承________类。
2. 绑定服务时必须实现的方法是________。
3. Service 的实现方法是________和________。
4. 在清单文件中注册服务时应该使用的节点是________。

二、选择题

1. 对于 Service 生命周期的 onCreate()和 onStart()，说法正确的是(　　)(多选)。
   A. 当第一次启动的时候先后调用 onCreate()和 onStart()方法
   B. 当第一次启动的时候只会调用 onCreate()方法
   C. 如果 Service 已经启动，将先后调用 onCreate()和 onStart()方法
   D. 如果 Service 已经启动，只会执行 onStart()方法，不再执行 onCreate()方法
2. 每一次启动服务时都会调用的方法是(　　)。
   A. onCreate()　　B. onStart()
   C. onResume()　　D. onStartCommand()
3. 下列不属于绑定服务特点的是(　　)。
   A. 以 bindService()方法开启　　B. 调用者关闭后服务关闭
   C. 必须实现 ServiceConnection()　　D. 使用 stopService()方法关闭服务
4. 下列不属于 Service 生命周期的方法是(　　)。
   A. onResume()　　B. onStart()　　C. onStop()　　D. onDestoty()

三、简答题

1. 什么时候使用 Service?
2. 如何在启动一个 Activity 时就启动一个 Service?

第7章 数据存储

本章重点

- 文件存储
- SharedPreferences
- SQLite 的基本操作
- 内容提供者

Android 操作系统为数据存储主要提供了四种方式：

(1) 使用文件存储(File 存储)；

(2) 首选项存储(Preferences 存储)；

(3) 数据库存储(SQLite 存储)；

(4) 内容提供者存储(Content Providers 存储)。

7.1 文件存储

默认情况下，Android 系统文件存储(File 存储)属于应用程序私有的，用户或其他应用程序都是无法直接访问的。当应用程序被卸载时，这些数据也会从内部存储中被清除。

Context 提供的 openFileOutput(String name, int mode)方法可以打开与此应用程序包相关联的私有文件的输出流。如果文件不存在，则创建文件。

1. 保存文件的示例代码

```
String FILE_NAME = "hello_file";
String string = "文件存储!";
FileOutputStream fos = openFileOutput(FILE_NAME, Context.MODE_PRIVATE);
fos.write(string.getBytes());
fos.close();
```

2. 完整代码(FileActivity .java)

```
public class FileActivity extends Activity {
EditText mFileName, mFileContent;
Button mWriteData, mReadData;
TextView mShow;
```

```
@Override
protected void onCreate(Bundle savedInstanceState) {
    super.onCreate(savedInstanceState);
    setContentView(R.layout.activity_file);
    initView();

}

/***
 * 初始化控件
 */
private void initView() {
    mFileName = (EditText) findViewById(R.id.edt_file_name);
    mFileContent = (EditText) findViewById(R.id.edt_file_content);
    mWriteData = (Button) findViewById(R.id.btn_write);
    mReadData = (Button) findViewById(R.id.btn_read);
    mShow = (TextView) findViewById(R.id.tv_read_content);
    mWriteData.setOnClickListener(new View.OnClickListener() {
        @Override
        public void onClick(View view) {
            if (mFileName.getText().toString() == "" && mFileContent.getText()
              .toString() == "") {
                Toast.makeText(FileActivity.this, "文件名或文件内容不能为空",
                  Toast.LENGTH_LONG).show();
            } else {
                String filename = mFileName.getText().toString();
                String filecontent = mFileContent.getText().toString();
                FileOutputStream fileOutputStream = null;
                try {
                    fileOutputStream = openFileOutput(filename, Context.
                      MODE_PRIVATE);
                    fileOutputStream.write(filecontent.getBytes("UTF-8"));
                } catch (Exception e) {
                    e.printStackTrace();
                } finally {
                    try {
                        fileOutputStream.close();
                    } catch (Exception e) {
                        e.printStackTrace();
                    }
                }
            }
        }
    });
    mReadData.setOnClickListener(new View.OnClickListener() {
        @Override
        public void onClick(View view) {
            String filename = mFileName.getText().toString();
            //获得欲读取文件的名称
            FileInputStream in = null;
            ByteArrayOutputStream bout = null;
```

```
                byte[]buf = new byte[1024];
                bout = new ByteArrayOutputStream();
                int length = 0;
                try {
                    in = openFileInput(filename);          //获得输入流
                    while((length = in.read(buf))!= -1){
                        bout.write(buf,0,length);
                    }
                    byte[] content = bout.toByteArray();
                    mShow.setText(new String(content,"UTF-8"));
                    //设置文本框为读取的内容
                } catch (Exception e) {
                    e.printStackTrace();
                }
                mShow.invalidate();                        //刷新屏幕
                try{
                    in.close();
                    bout.close();
                }
                catch(Exception e){}
            }
        });
    }
}
```

3. 布局文件(activity_file.xml)代码

```
<?xml version = "1.0" encoding = "utf-8"?>
<LinearLayout xmlns:android = "http://schemas.android.com/apk/res/android"
xmlns:tools = "http://schemas.android.com/tools"
android:layout_width = "match_parent"
android:layout_height = "match_parent"
android:background = "#fff"
android:orientation = "vertical"
android:padding = "5dp">

<TextView
    android:layout_width = "wrap_content"
    android:layout_height = "wrap_content"
    android:layout_marginTop = "10dp"
    android:text = "文件名(带后缀):"
    android:textSize = "16sp" />

<EditText
    android:id = "@+id/edt_file_name"
    android:layout_width = "match_parent"
    android:layout_height = "wrap_content"
    android:layout_marginBottom = "10dp"
    android:layout_marginTop = "10dp"
    android:hint = "请输入文件名+后缀名"
    android:padding = "5dp"
```

```
        android:textSize="14sp" />

    <TextView
        android:layout_width="wrap_content"
        android:layout_height="wrap_content"
        android:text="文件内容: "
        android:textSize="16sp" />

    <EditText
        android:id="@+id/edt_file_content"
        android:layout_width="match_parent"
        android:layout_height="wrap_content"
        android:layout_marginBottom="10dp"
        android:layout_marginTop="10dp"
        android:hint="请输入文件名+后缀名"
        android:padding="5dp"
        android:textSize="14sp" />

    <LinearLayout
        android:layout_width="wrap_content"
        android:layout_height="wrap_content"
        android:layout_marginBottom="10dp"
        android:orientation="horizontal">

        <Button
            android:id="@+id/btn_write"
            android:layout_width="wrap_content"
            android:layout_height="wrap_content"
            android:padding="10dp"
            android:text="写入数据" />
        <Button
            android:id="@+id/btn_read"
            android:layout_width="wrap_content"
            android:layout_height="wrap_content"
            android:padding="10dp"
            android:layout_marginLeft="20dp"
            android:text="读取数据" />

    </LinearLayout>
    <TextView
        android:id="@+id/tv_read_content"
        android:layout_width="match_parent"
        android:hint="显示内容文本框"
        android:padding="5dp"
        android:layout_height="wrap_content" />
</LinearLayout>
```

运行 FileActivity.java，界面如图 7-1 所示。打开 DDMS 的“File Explorer”选项卡，在 /data/data/<packagename>/files 目录下出现了 test.txt 文件，如图 7-2 所示。单击工具栏按钮“pull a file from the device”可以将数据文件 test.txt 导出，如图 7-3 所示。

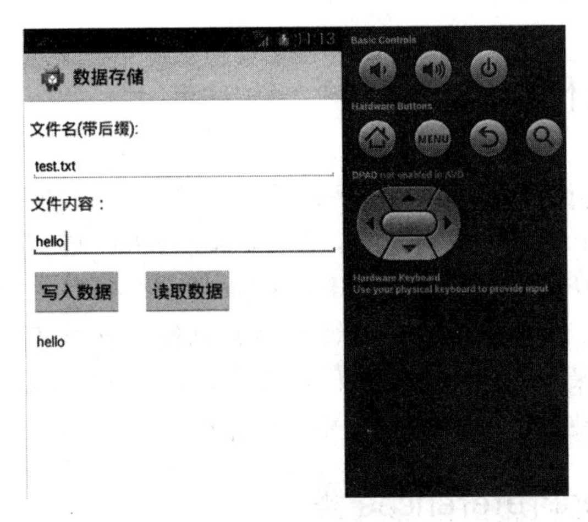

图 7-1 AVD 截屏

Problems | Javadoc | Declaration | Console | Progress | LogCat | PyUnit | Remote Systems | File Explorer

Name	Size	Date	Time	Permissions	Info
com.qsd.ch7_1		2017-10-07	02:35	drwxrwx--x	
cache		2017-09-13	08:13	drwxrwx--x	
files		2017-10-07	02:40	drwxrwx--x	
test.txt	5	2017-10-07	02:40	-rw-rw----	
lib		2017-10-07	02:35	lrwxrwxrwx	-> /data/a...
com.svox.pico		2017-06-23	21:39	drwxr-x--x	

图 7-2 文件存储位置

com.qsd.ch7_1		2017-02-24	11:07	drwxr-x--x	
cache		2017-02-24	11:05	drwxrwx--x	
files		2017-02-24	11:07	drwxrwx--x	
test.txt	6	2017-02-24	11:07	-rw-rw----	
lib		2017-02-24	11:04	lrwxrwxrwx	-> /data/a...
com.svox.pico		2016-09-04	14:33	drwxr-x--x	
com.wandoujia.phoenix2		2017-02-24	11:03	drwxr-x--x	

图 7-3 通过 DDMS 查看文件

Context 提供的 openFileIntput(String name, int mode)方法可以打开与此应用程序包相关联的私有文件的输入流，将数据从设备的文件中读出。

openFileIntput(String name, int mode)方法的第二个参数主要有四种模式，如表 7-1 所示。

表 7-1 openFileIntput()方法的四种模式

模 式	功 能 描 述
Context. MODE_PRIVATE	私有模式，也是默认模式。由其创建的文件只能由调用应用程序(或共享相同用户 ID 的所有应用程序)访问
Context. MODE_APPEND	追加模式。如果文件已经存在，则写数据到现有文件的末尾而不是消除它
Context. MODE_WORLD_ READABLE	可读模式。允许所有其他应用程序对已创建的文件进行读访问
Context. MODE_WORLD_ WRITABLE	可写模式，允许所有其他应用程序对已创建的文件进行写访问

7.2 首选项存储

首选项存储(Preference 存储)主要是存储一些应用程序的配置信息,如用户名、口令及自定义参数等。Preference 采用"键-值"对(key-value)格式组织和管理数据,其数据存储在 XML 文件中,该文件在 data/data/< packagename >/shared_prefs 目录下。相对于其他方式,首选项存储是一个轻量级的存储机制。该方式实现比较简单,可以使用 SharedPreferences 保存诸如 booleans,floats,ints,longs 和 strings 的任何原始数据。此数据将在用户会话(即使应用程序被杀死)中持久存在。

7.2.1 SharedPreferences 类

SharedPreferences 是 Android 平台上一个轻量级的存储类,使用 SharedPreferences 存储数据的步骤如下:

(1) 使用 getSharedPreferences(string name,int mode)生成 SharedPreferences 对象。示例代码如下:

```
SharedPreferences sharedPreferences = context.getSharedPreferences(string name, int mode);
```

其中,参数 name 表示存储数据的 XML 文件名,文件名不需要指定.xml 后缀;参数 mode 表示文件操作模式,有 MODE_WORLD_READBLE(可读)、MODE_WORLD_WRITEABLE(可写)、MODE_PRIVATE(私有)、MODE_APPEND(可追加)多种选择。

(2) 使用 SharedPreferences. Editor 的 putXXX()方法保存数据。

(3) 使用 SharedPreferences. Editor 的 commit()方法将上一步得到的数据保存到 XML 文件中。

(4) 使用 SharedPreferences 的 getXXX()方法获取数据。

其中,putXXX()、getXXX()中的 XXX 会随数据类型是 string、float、int 或者 long 等类型的变化而相应变化。

使用 Preferences 存取数据时,需要用到 android. content 包中的两个对象:SharedPreferences 和 SharedPreferences. Editor,它们提供了许多方法用于获取数据、存储和修改数据,具体如表 7-2、表 7-3 所示。

表 7-2 SharedPreferences 的常用方法

方　法	功能描述
contains (string key)	判断是否包含特定 key 的数据
edit()	返回 SharedPreferencesEditor 对象
getAll ()	获取全部的键值对
getBoolean (string key, boolean defValue)	获取指定 key 的布尔值
getFloat (string key, float defValue)	获取指定 key 的 float 值
getInt (string key, int defValue)	获取指定 key 的 int 值
getString (string key, string defValue)	获取指定 key 的 string 值
getLong (string key, long defValue)	获取指定 key 的 long 值

表 7-3 SharedPreferences. Editor 的常用方法

方法	功能描述
clear()	清除 SharedPreferences 所有值
commit()	编辑结束后,提交数据
putBoolean(string key, boolean value)	保存指定 key 的 boolean 值
putFloat(string key, float value)	保存指定 key 的 float 值
putInt(string key,int value)	保存指定 key 的 int 值
putLong(string key, long value)	保存指定 key 的 long 值
putString(string key, string value)	保存指定 key 的 string 值

7.2.2 使用 Preference 存储的案例——简单登录界面

本例实现登录时保存账号和密码,在登录完成之后打开登录界面,用户名和密码自动填写完成。布局文件的 xml 代码如下:

```
<?xml version = "1.0" encoding = "utf - 8"?>
<LinearLayout xmlns:android = "http://schemas.android.com/apk/res/android"
    xmlns:tools = "http://schemas.android.com/tools"
    android:layout_width = "match_parent"
    android:layout_height = "match_parent"
    android:background = "#fff"
    android:orientation = "vertical"
    android:padding = "5dp">

    <TextView
        android:layout_width = "wrap_content"
        android:layout_height = "wrap_content"
        android:layout_marginTop = "10dp"
        android:text = "用户名: "
        android:textSize = "16sp" />

    <EditText
        android:id = "@ + id/edt_name"
        android:layout_width = "match_parent"
        android:layout_height = "wrap_content"
        android:layout_marginBottom = "10dp"
        android:layout_marginTop = "10dp"
        android:hint = "请输入用户名"
        android:padding = "5dp"
        android:textSize = "14sp" />

    <TextView
        android:layout_width = "wrap_content"
        android:layout_height = "wrap_content"
        android:text = "密码: "
        android:textSize = "16sp" />

    <EditText
```

```
        android:id="@+id/edt_pass"
        android:layout_width="match_parent"
        android:layout_height="wrap_content"
        android:layout_marginBottom="10dp"
        android:layout_marginTop="10dp"
        android:hint="请输入密码"
        android:inputType="textPassword"
        android:padding="5dp"
        android:textSize="14sp" />

    <CheckBox
        android:id="@+id/cb_save"
        android:layout_width="wrap_content"
        android:layout_height="wrap_content"
        android:layout_marginBottom="10dp"
        android:layout_marginTop="10dp"
        android:checked="true"
        android:text="保存账号" />

    <Button
        android:id="@+id/btn_login"
        android:layout_width="wrap_content"
        android:layout_height="wrap_content"
        android:padding="10dp"
        android:text="登录" />


</LinearLayout>
```

LoginActivity 类主要代码如下：

```
package com.qsd.ch7_2_2;

import android.app.Activity;
import android.content.SharedPreferences;
import android.os.Bundle;
import android.util.Log;
import android.view.View;
import android.widget.Button;
import android.widget.CheckBox;
import android.widget.EditText;
import android.widget.Toast;

import com.qsd.ch7_2.R;

public class LoginActivity extends Activity {
    EditText mName, mPass;
    Button mLogin;
    CheckBox mcb;
    SharedPreferences sharedPreferences;
```

```
@Override
protected void onCreate(Bundle savedInstanceState) {
    super.onCreate(savedInstanceState);
    setContentView(R.layout.activity_login);
    initView();
    initData();
}

private void initData() {
    sharedPreferences = getSharedPreferences("user", MODE_PRIVATE);
}

@Override
protected void onStart() {
    super.onStart();
    String name = sharedPreferences.getString("name", "").toString();
    String pass = sharedPreferences.getString("pass", "").toString();
    if (!name.equals("")) {// 不为空时
        mName.setText(name);
        mPass.setText(pass);
    } else {
        Log.e("test", "test");
    }
}

private void initView() {
    mName = (EditText) findViewById(R.id.edt_name);
    mPass = (EditText) findViewById(R.id.edt_pass);
    mLogin = (Button) findViewById(R.id.btn_login);
    mcb = (CheckBox) findViewById(R.id.cb_save);
    mLogin.setOnClickListener(new View.OnClickListener() {
        @Override
        public void onClick(View view) {
            if (mName.getText().toString() == ""
                    && mPass.getText().toString() == "") {
                Toast.makeText(LoginActivity.this, "用户名或密码不能为空",
                    Toast.LENGTH_LONG).show();
            } else {
                if (mcb.isChecked()) {// 如果选中,则保存账号和密码

                    SharedPreferences.Editor editor = sharedPreferences
                        .edit();
                    editor.putString("name", mName.getText().toString());
                    editor.putString("pass", mPass.getText().toString());
                    editor.commit();
                }
                finish();
            }
```

```
            }
        });
    }

}
```

在登录界面，输入用户名及密码，单击“登录”按钮，效果如图 7-4 所示。

图 7-4 运行效果

打开 DDMS 的“File Explorer”选项卡，在/data/data/< packagename >/shared_prefs 目录下出现了 user. xml 文件，如图 7-5 所示。单击 DDMS 工具栏按钮“pull a file from the device”可以将数据文件 user. xml 导出。

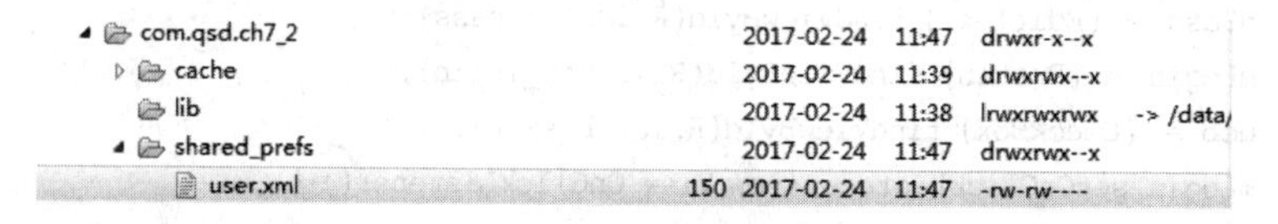

图 7-5 数据存储位置

7.3 SQLite 存储

SQLite 使用数据库作为存储方式，它是一个重量级的存储机制，适合大数据量的数据存储(如联系人列表或者数据库)。通过这种方式能够很容易地对数据进行增加、插入、删除、更新等操作。相比首选项存储和文件存储，使用 SQLite 较为复杂。

可以为应用程序创建一个私有数据库，应用程序的各个组件均可以访问这些数据，而其他的应用程序在未获得授权时无法访问。

7.3.1 SQLiteOpenHelper 类

SQLiteOpenHelper 是一个抽象类，用于创建数据库和数据库版本的更新。创建一个 SQLite 数据库，需要创建一个 SQLiteOpenHelper 类的子类，并重写其中的 onCreate()方法。SQLiteOpenHelper 的常用方法如表 7-4 所示。

表 7-4 SQLiteOpenHelper 的常用方法

方 法	功能描述
SQLiteOpenHelper(Context context, String name, SQLiteDatabase. CursorFactory, int version)	构造函数。第 1 个参数是上下文对象；第 2 个参数是数据库名称；第 3 个参数是游标工厂；第 4 个参数是版本号,最小值为 1
onCreate(SQLiteDatabase db)	创建数据库时调用
onUpgrade(SQLiteDatabase db, int oldVersion,int newVersion)	数据库版本更新时调用
getReadableDatabase()	创建或打开一个只读数据库
getWritableDatabase()	创建或打开一个读写数据库

7.3.2 SQLiteDatabase 类

SQLiteDatabase 是一个数据库访问类,它封装了许多数据库操作的 API,利用这些 API 可以对数据库进行增、删、改、查的操作。SQLiteDatabase 操作数据库的常用方法如表 7-5 所示。

表 7-5 SQLiteDatabase 的常用方法

方 法	功能描述
openOrCreateDatabase(String path,SQLiteDatabase. CursorFactory factory)	打开或创建一个数据库
openDatabase(String path, SQLiteDatabase. CursorFactory factory, int flags)	打开指定的数据库
insert(String table,String nullColumnHack,ContentValues values)	添加一条记录
query (boolean distinct, String table, String[] columns, String selection, String[] selectionArgs, String groupBy, String having, String orderBy, String limit)	查询数据
delete(String table,String whereClause,String[] whereArgs)	删除一条记录
update (String table, ContentValues value, String whereClause, String [ ] whereArgs)	修改记录
execSQL(String sql)	执行一条 SQL 语句
close()	关闭数据库

7.3.3 Cursor 游标

Cursor 是一个游标接口,它是一个满足某种查询条件的数据结果集,驻留在内存中。Cursor 游标的一些常见方法如表 7-6 所示。

表 7-6 Cursor 的常用方法

方 法	功能描述
moveToFisrt()	将游标移动到第一行
moveToLast()	将游标移动到最后一行
moveToPrevious()	移动游标到上一行
moveToNext()	将游标移动到下一行

续表

方　　法	功能描述
moveToPosition(int position)	将游标移动到一个绝对的位置
getCount()	返回游标中的行数
getPosition ()	返回当前游标的位置
getColumnName(int columnIndex)	从给定的索引返回列名
getColumnNames()	返回一个字符串数组的列名
close()	关闭游标，释放资源

以上介绍了 Android SDK 提供的操作 SQLite 数据库所涉及的一些类及其常用的方法。熟练掌握它们的使用可以有效地操作 SQLite 数据库。

7.3.4 SQLite 数据库操作方法

1. 创建 SQLite 数据库

Android 推荐使用 SQLiteOpenHelper 的子类创建 SQLite 数据库，具体代码如下：

```
public class MySQLiteOpenHelper extends SQLiteOpenHelper {
   //数据库的构造方法
    public MySQLiteOpenHelper(Context context, String name,
            CursorFactory factory, int version) {
        super(context, "student.db", null, 6);
            }
    //数据库第一次被创建时调用该方法
    public void onCreate(SQLiteDatabase db) {
        // 创建一个数据库表的 SQL 命令
        db.execSQL("create table student(id integer primary key autoincrement," +
         "name varchar(10)," + "grade number(3))");
    }
    //当 arg2 > arg1 时，升级数据库

    public void onUpgrade(SQLiteDatabase db, int arg1, int arg2) {
        // TODO Auto - generated method stub
    }
}
```

说明：创建的数据库 student.db 存放在/data/data/packagename/database 目录下。

2. 插入一条记录

添加数据可以使用 SQLiteDatabase.execSQL(String sql, Object[] bindArgs)方法来实现，具体如下：

```
public void addStudentInfo(Student student) {
        db = mySQLiteOpenHelper.getWritableDatabase();
        db.execSQL("INSERT INTO tab_student (id, name, grade) values (?, ?, ?)",
          new Object[] {student.getId(), student.getName(),
student.getGrade()});
    }
```

其中,通过第二个参数 bindArgs 使 SQL 语句中的问号(?)与数组 Object[]中的值形成一一对应关系,从而将值写入到 student 表中的对应字段中。

3. 修改一条记录

修改记录的方法与插入记录的方法类似,具体如下:

```
public void updateStudentInfo(Student student) {
 db = mySQLiteOpenHelper.getWritableDatabase();
      db.execSQL("UPDATE student SET Name = ?, Grade = ? WHERE Id = ?",
        new Object[] {student.getName(),student.getGrade(),
student.getId()});
  }
```

4. 查询记录

查询记录时,因为需要返回查询的结果,所以需要使用 SQLiteDatabase.rawQuery()方法将查询的结果返回,具体如下:

```
public Student findStudentInfo(int id) {
        db = mySQLiteOpenHelper.getWritableDatabase();
         String sql = "SELECT Id, Name, Grade FROM student WHERE Id = ?";
        Cursor cursor = db.rawQuery(sql, new String[] {String.valueOf(id)});
         if(cursor.moveToNext()) {
            return new Student(cursor.getInt(cursor.getColumnIndex("Id")),
                cursor.getString(cursor.getColumnIndex("Name")),cursor.getInt
                (cursor.getColumnIndex("Grade")));
         }
         return null;
    }
```

7.3.5 使用 SQLite 存储的案例——歌曲列表浏览

本例可实现歌曲的添加与存储,在歌曲添加完成之后打开歌曲界面,自动展示所添加歌曲的所有信息。AddMusic 代码如下:

```
public class AddMusic extends Activity {
    private EditText edt1, edt2;
    private Button btn1;
    @Override
    public void onCreate(Bundle savedInstanceState) {
        super.onCreate(savedInstanceState);
        setContentView(R.layout.main);
        this.setTitle("添加信息");
        edt1 = (EditText) findViewById(R.id.edt_test1);
        edt2 = (EditText) findViewById(R.id.edt_test2);
        btn1 = (Button) findViewById(R.id.btn_test);
        btn1.setOnClickListener(new OnClickListener() {
            public void onClick(View v) {
                // 输入信息
```

```
                String musicname = edt1.getText().toString();
                String singername = edt2.getText().toString();
                // 封装信息
                ContentValues values = new ContentValues();
                values.put("musicname", musicname);
                values.put("singername", singername);
                // 创建 DBHelper 类
                MyDB helper = new MyDB(getApplicationContext());
                //插入数据
                helper.insert(values);
                // 跳转到 Search,显示歌曲清单
                Intent myintent = new Intent(AddMusic.this,
                        Search.class);
                startActivity(myintent);
            }
        });
    }
}
```

Search 代码如下：

```
public class Search extends ListActivity {
    @Override
    public void onCreate(Bundle savedInstanceState) {
        super.onCreate(savedInstanceState);
        this.setTitle("浏览歌曲清单");
        final MyDB helpter = new MyDB(this);
        // 声明游标
        Cursor cursor = helpter.query();
        // 声明数组
        String[] from = { "number", "musicname", "singername" };
        int[] to = { R.id.tv_test01, R.id.tv_test02, R.id.tv_test03 };
        // 声明适配器
        SimpleCursorAdapter adapter = new SimpleCursorAdapter(this,
            R.layout.list, cursor, from, to);
        //列表视图
        ListView listView = getListView();
        //添加适配器
        listView.setAdapter(adapter);
        //提示对话框
        final AlertDialog.Builder builder = new AlertDialog.Builder(this);
        //设置 ListView 监听器
        listView.setOnItemClickListener(new OnItemClickListener() {
            @Override
            public void onItemClick(AdapterView<?> arg0, View arg1, int arg2,
                long arg3) {
                final long temp = arg3;
                builder.setMessage("确定删除吗?").setPositivePositiveButton("是",
                        new DialogInterface.OnClickListener() {
                            public void onClick(DialogInterface dialog,
                                    int which) {
```

```
                            //删除数据
                            helpter.del((int) temp);
                            //重新查询数据
                            Cursor c = helpter.query();
                            String[] from = { "number", "musicname",
                                    "singername" };
                            int[] to = { R.id.tv_test01, R.id.tv_test02,
                                R.id.tv_test03 };
                            SimpleCursorAdapter adapter = new
                                SimpleCursorAdapter(getApplicationContext(),
                                R.layout.list, c, from, to);
                            ListView listView = getListView();
                            listView.setAdapter(adapter);
                        }
                    }).setNegativeButton("否",
                    new DialogInterface.OnClickListener() {
                        public void onClick(DialogInterface dialog,
                                int which) {
                        }
                    });
                AlertDialog alertdialog = builder.create();
                alertdialog.show();
            }
        });
        helpter.close();
    }
}
```

MyDB 代码如下：

```
public class MyDB extends SQLiteOpenHelper {
    // 数据库名称
    private static final String DB_NAME = "music.db";
    // 数据表名
    private static final String TBL_NAME = "MusicTbl";
    // 声明 SQLiteDatabase 对象
    private SQLiteDatabase db;
    // 构造函数
    MyDB(Context c) {
        super(c, DB_NAME, null, 2);
    }

    @Override
    public void onCreate(SQLiteDatabase db) {
        // 获取 SQLiteDatabase 对象
        this.db = db;
        // 创建表
        String CREATE_TBL = " create table "
            + " MusicTbl(number integer primary key autoincrement,musicname
              text,singername text) ";
        db.execSQL(CREATE_TBL);
```

```
    }
    // 插入
    public void insert(ContentValues values) {
        SQLiteDatabase db = getWritableDatabase();
        db.insert(TBL_NAME, null, values);
        db.close();
    }
    // 查询
    public Cursor query() {
        SQLiteDatabase db = getWritableDatabase();
        Cursor c = db.query(TBL_NAME, null, null, null, null, null, null);
        return c;
    }
    // 删除
    public void del(int id) {
        if (db == null)
            db = getWritableDatabase();
        db.delete(TBL_NAME, "_id = ?", new String[] { String.valueOf(id) });
    }
    // 关闭数据库
    public void close() {
        if (db != null)
            db.close();
    }
    @Override
    public void onUpgrade(SQLiteDatabase db, int oldVersion, int newVersion) {
    }
}
```

运行程序，效果如图 7-6 所示。

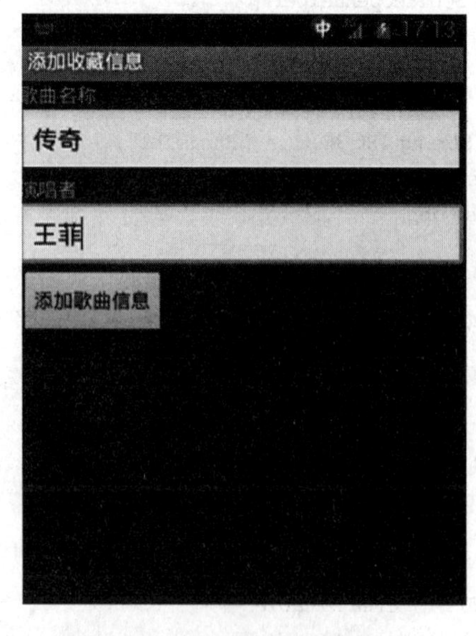

图 7-6 运行效果

打开 DDMS 的“File Explorer”选项卡，在/data/data/< packagename >/databases 目录下出现了 music. db 文件。单击 DDMS 工具栏按钮“pull a file from the device”可以将数据库文件 music. db 导出。文件 music. db-journal 是 sqlite 的一个临时的日志文件，主要用于 sqlite 事务回滚机制，在事务开始时产生。如图 7-7 所示。

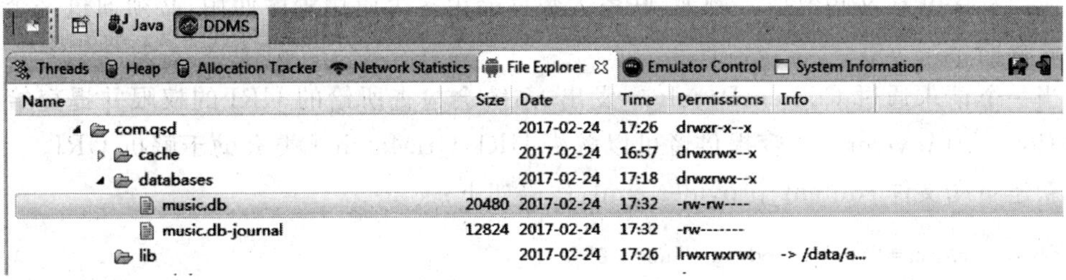

图 7-7 数据文件位置

利用第三方工具软件 SQLite Expert Personal 打开从 DDMS 的“File Explorer”选项卡导出的数据库，如图 7-8 所示，可利用关系数据库 SQL 命令操作数据库。

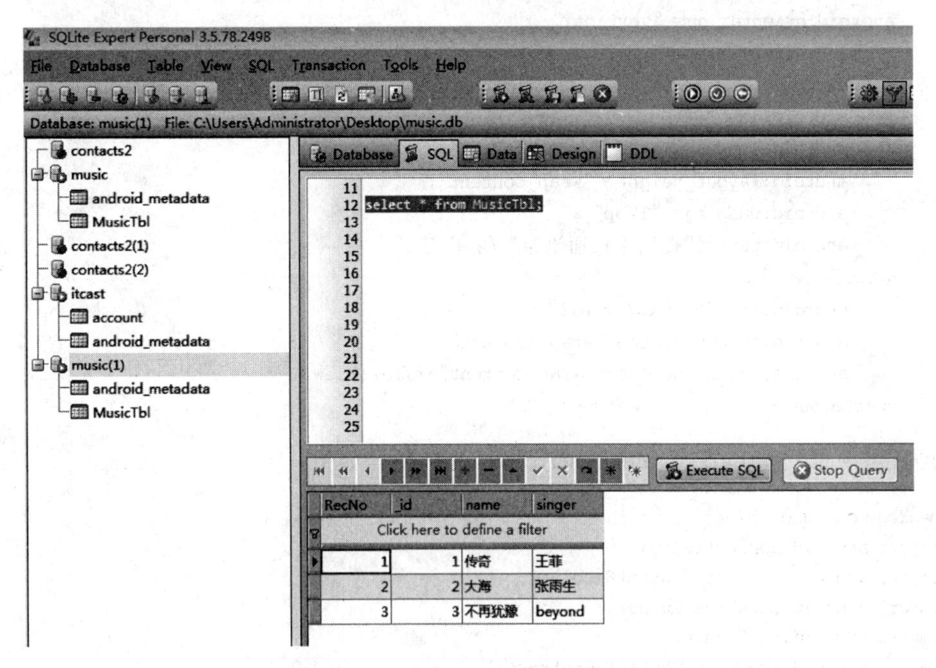

图 7-8 SQLite Expert Personal 运行效果

7.4 内容提供者存储

在 Android 平台中，没有提供一个公共数据存储区域供所有应用程序共享数据。在 7.1 节曾提到，通过将文件输出流的 openFileOutput(FILE_NAME, Context. MODE_PRIVATE)方法的第二个参数 mode 修改为 Context. MODE_WORLD_READABLE 和 Context. MODE_WORLD_WRITEABLE，可以让其他应用程序访问该应用程序的数据文件。但这种方法的缺点是，需要知道文件的存储目录，且文件内容也会暴露。Android 平台提供了内容提供者(Content Providers)存储，实现了应用程序间安全的数据共享。

内容提供者存储是 Android 应用程序的主要构件之一，为应用程序提供内容。内容提供者封装数据并通过单一 ContentResolver 接口提供给应用程序。内容提供者只需知道有多少应用程序共享这些数据。例如，联系人数据会由多个应用程序使用，必须通过一个内容提供者存储。

当一个请求通过 ContentResolver 发出，系统会检查所给的 URI 的权限并提交给已注册的 ContentProvider。内容提供者可以解析 URI，UriMatcher 类有助于解析 URI。

新建布局文件 content_provider.xml，代码如下：

```
<?xml version="1.0" encoding="utf-8"?>
<LinearLayout xmlns:android=
  "http://schemas.android.com/apk/res/android"
    xmlns:tools="http://schemas.android.com/tools"
    android:layout_width="match_parent"
    android:layout_height="match_parent"
    android:orientation="vertical"
    >
    <Button
        android:id="@+id/btn_query"
        android:layout_width="wrap_content"
        android:layout_height="wrap_content"
        android:padding="12dp"
        android:text="查询手机通讯录" />
    <ListView
        android:id="@+id/lv_all"
        android:layout_width="wrap_content"
        android:layout_height="wrap_content"></ListView>
</LinearLayout>
```

再新建一个 ContentProviderTestActivity 类，代码如下：

```
package com.qsd.ch7_4;
import android.app.Activity;
import android.content.ContentResolver;
import android.database.Cursor;
import android.os.Bundle;
import android.provider.ContactsContract;
import android.view.View;
import android.widget.ArrayAdapter;
import android.widget.Button;
import android.widget.ListView;
import com.qsd.ch7_4.R;
public class ContentProviderActivity extends Activity {
    ListView mylv;
    Button mybtn;
    @Override
    protected void onCreate(Bundle savedInstanceState) {
        super.onCreate(savedInstanceState);
        setContentView(R.layout.content_provider);
```

```
        initVew();
    }
    private void initVew() {
        mylv = (ListView) findViewById(R.id.mylv);
        mybtn = (Button) findViewById(R.id.mybtn);
        mybtn.setOnClickListener(new View.OnClickListener() {
            @Override
            public void onClick(View view) {
                queryContact();
            }
        });
    }
    // 查询手机通讯录
    private void queryContact() {
        ContentResolver cr = getContentResolver();
        Cursor cs = cr.query(ContactsContract.Contacts.CONTENT_URI,
                new String[] { ContactsContract.Contacts._ID,
                    ContactsContract.Contacts.DISPLAY_NAME }, null, null,
                    null);
        StringBuffer sBuffer = new StringBuffer();
        if (cs != null) {
            while (cs.moveToNext()) {
                // 先获取联系人的 id
                int id = cs.getInt(cs
                        .getColumnIndex(ContactsContract.Contacts._ID));
                            String name = cs.getString(cs.getColumnIndex(
                            ContactsContract.Contacts.DISPLAY_NAME));
                sBuffer.append(name + ",");
            }
            // 关闭游标
            cs.close();
            String[] items = sBuffer.toString().split(",");
            mlv.setAdapter(new ArrayAdapter<String>(this,
             android.R.layout.simple_list_item_1, items));
        }
    }
}
```

单击“查询手机通讯录”按钮，就会看到事先录入的手机通讯录的联系人信息。运行效果如图 7-9 所示。

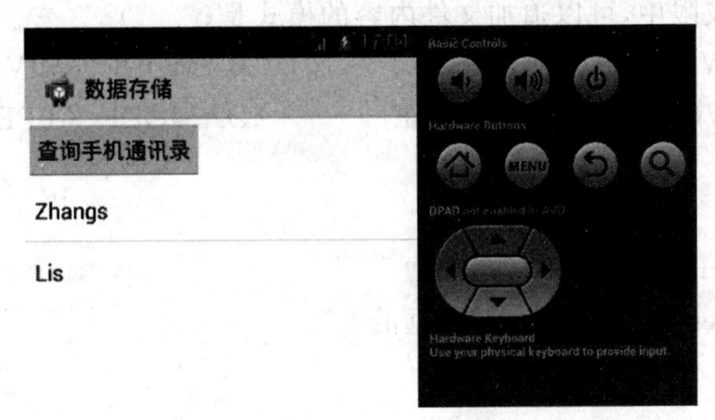

图 7-9 查询手机通讯录

获取联系人也是需要权限的,在清单文件 AndroidManifest. xml 中添加如下权限:

```
<uses-permission android:name="android.permission.WRITE_CONTACTS" />
<uses-permission android:name="android.permission.READ_CONTACTS" />
```

7.5 本章小结

本章介绍了 Android 四种数据存储方式。文件(File)存储方式通过 Context. openFileInput()方法获取标准的文件输入流(FileInputStream),读取设备上的文件,通过 Context. openFileOuput()方法获取标准的文件输出流(FileOutputStream)。首选项(Preferences)存储方式提供了一种轻量级的数据存储方式,以"key-value"方式将数据保存在一个 XML 清单文件中。SQLite 存储方式实现了结构化数据存储,SQLiteOpenHelper 是 SQLiteDatabase 的一个帮助类,用来管理数据库的创建和版本的更新。内容提供者(Content Providers)存储方式,是应用程序之间数据存储和检索的一个渠道,其作用就是使得各个应用程序之间实现数据共享,Android 系统为一些常见的应用(如联系人列表、音乐、视频、图像等)定义了相应的 ContentProvider,它们被定义在 android. provider 包下。除了所介绍的四种存储方式外,Android 数据存储还可以采用网络存储技术(NetWork),读者可以参阅其他相关资料。

7.6 练习题

一、填空题

1. Android 中的文件可以存储在________和________中。
2. SharedPreferences 是一种轻量级的数据存储方式,主要用来存储应用程序的________。

二、选择题

1. 应用程序共享数据的存储方式是(　　)。
   A. SQLite 存储　　B. File 存储
   C. Preference 存储　　D. Content Providers 存储
2. 在文件操作权限中,可以追加文件内容的模式是(　　)。
   A. MODE_WORLD_READABLE　　B. MODE_PRIVATE
   C. MODE_WORLD_WRITEABLE　　D. MODE_APPEND

三、简答题

1. SQLiteOpenHelper 的作用是什么?
2. SharedPreferences 是如何存储数据的?

四、编程题

使用 SQLite 数据存储方式,编写一个学生信息展示程序。

第8章 网络通信

本章重点

- 掌握 Android 网络通信原理
- 掌握 Socket、HTTP、URL 以及 WebView 等网络通信机制
- 了解 Handle 线程间通信

伴随着无线网络通信技术的不断升级换代，诸如无线宽带上网、无线搜索、视频聊天、网络游戏等得到快速普及与发展。因此，研究移动通信设备网络通信技术，开发出优质的 Android 网络应用程序，有着很好的应用前景。

Android 是由 Google 牵头开发的。Google 的应用层采用 Java 语言，所以 Java 支持的网络编程模式在 Android 中都能支持。

Android 中常用的网络编程方式如下：

- 针对 TCP/IP，使用 Socket(套接字)和 ServerSocket；
- 针对 HTTP，使用 HttpURLConnection 和 HttpClient；
- 直接使用 WebView 访问网络。

8.1 Socket 通信

通过 Socket 实现 Android 与服务器的通信是一种常用的通信方式。Socket 通常被称为"套接字"，它是支持 TCP/IP 网络通信的基本操作单元，包括网络通信涉及的五种信息：通信协议、本地端口、本地 IP 地址、远程端口以及远程 IP 地址。跨平台的 Socket 编程方式可以实现异构系统之间的网络通信。

java.net 包中提供 Socket 和 ServerSocket 表示双向连接的 Client 和 Server。其中，java.net.Socket 是客户端的 Socket 对应的类，java.net.ServerSocket 是服务器端的 Socket 对应的类。java.net.ServerSocket 类包含一个等待客户端连接的服务器端套接字，即 IP 地址和端口号 PORT。

通过套接字实现通信一般分为三步：首先服务器监听，然后客户端请求，最后连接确认。对应于 Java 中的工作过程具体如下：

(1) 服务器监听。

(2) 客户端请求。

(3) 连接确认。

8.1.1 Socket 客户端的开发

Socket 类常用的方法见表 8-1。一个典型的创建客户端 Socket 的部分过程代码如下：

```
try{
Sockets = new Socket("219.218.22.133",4500);
}
catch(IOException ioe) {
System.out.println("Error:" + ioe);
}
```

表 8-1 Socket 类常用方法及功能

方　法	功能描述
Socket(InetAddreess address, int port) Socket(InetAddreess address, int port, boolean steam) Socket(String host,int port) Socket(String host, int port, boolean stream) Socket(SocketImp impl) Socket(String host, int port, InetAddreess localAddr, int localport) Socket(InetAddreess address, Int port, InetAddreess localAddr, int localport)	Socket 主要提供了 7 个构造方法,用于与服务器通信。其中 address 表示 IP 地址,host 表示主机名,port 表示端口号,stream 指明 Socket 是流还是数据报,localport 表示本地主机的端口号,localAddr 是本地机器的地址,impl 是 Socket 的父类,既可以用来创建 Socket 类,也可以用来创建 ServerSocket 类
bind(SocketAddress localAddr)	将该 Socket 同参数 localAddr 指定的地址和端口绑定
InetAddress getInetAddress()	获取该 Socket 连接的服务器的 IP 地址
synchronized int getReceiveBufferSize()	获取该 Socket 的接收缓冲区的尺寸
InputStream getInputStream()	获取该 Socket 的输入流,这个输入流用来读取数据
boolean isConnected()	判断该 Socket 是否连接
boolean isOutputShutdown()	判断该 Socket 的输出管道是否关闭
boolean isInputShutdown()	判断该 Socket 的输入管道是否关闭
SocketAddress getLocalSocketAddress()	获取此 Socket 的本地地址和端口
int getPort()	获取端口号
synchronized void close()	关闭 Socket

通信连接建立后,可以通过输入/输出流进行数据的传输。其中,服务器端的 OutputStream 为客户端的 InputStream 提供数据,同样,客户端的 OutputStream 为服务器端的 InputStream 提供数据。因此,在网络通信编程过程中,首先要处理好这些 I/O 之间的关系。

注意：网络通信结束后应及时关闭 Socket 和 Stream,释放占用的资源。虽然 Java 具有自动回收机制,但尚不足以彻底解决较为严重的资源拥塞,建议大家采取主动释放资源的方式。

通常 Socket 的工作步骤,参见图 8-1。

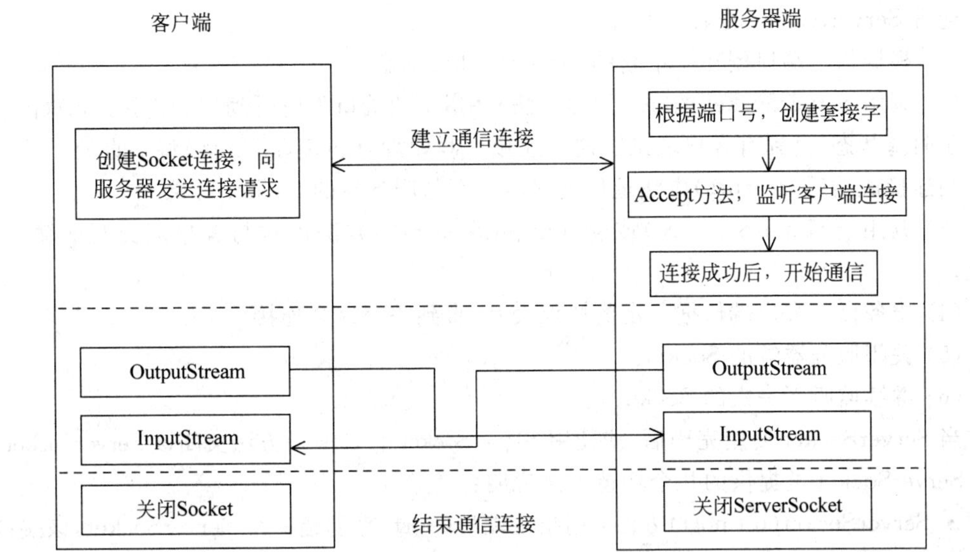

图 8-1 基于 Socket 套接字的通信机制

(1) 根据指定地址和端口创建一个 Socket 对象。

(2) 调用 getInputStream()方法或 getOutputStream()方法打开连接到 Socket 的输入/输出流。

(3) 客户端与服务器端依据一定的协议交互，直至关闭连接。

(4) 关闭客户端的 Socket。

8.1.2 Socket 服务器端的开发

ServerSocket 类用于服务器端的具体实现，它用于监听指定端口的 TCP 连接。当客户端的 Socket 试图与服务器端指定端口建立通信连接时，服务器会被激活并判定客户端的连接。当这个连接建立后，客户端与服务器端就可以相互传递数据，如图 8-1 所示。ServerSocket 类常用的方法如表 8-2 所示。

表 8-2 ServerSocket 类常用方法及功能

方　法	功能描述
ServerSocket(int port)	ServerSocket 提供了 3 个构造方法，用于实现服务器程序。
ServerSocket(int port, int backlog) ServerSocket(int port, int backlog, InetAddress bindAddr)	参数 port、backlog 和 bindAddr 分别代表连接中另一方的端口、连接请求的最大队列长度及本机地址
Socket accept()	等待客户端的连接，当客户端请求连接时，返回一个 Socket
SocketAddress getLocalSocketAddress()	获取此 Socket 的本地地址和端口
int getLocalPort()	获取端口号
InetAddress getInetAddress()	获取该 Socket 的 IP 地址
boolean isClosed()	判断连接是否关闭
void setSoTimeout(int timeout)	设置 accept 的超时时间
void close()	关闭服务器 Socket

通常 ServerSocket 的工作步骤：

(1) 根据指定端口创建一个新的 ServerSocket 对象。

(2) 调用 ServerSocket 的 accept()方法，在指定的端口中监听到来的连接。accept()一直处于阻塞状态，直到有客户端试图建立连接。假如没有连接，accept()将一直处于等待状态。当连接成功后，accept()方法返回连接客户端与服务器的 Socket 对象。

(3) 调用 getInputStream()或 getOutputStream()方法创建与客户端交互的输入/输出流。

(4) 服务器与客户端依据一定的协议交互，直到关闭这个连接。

(5) 关闭服务器端的 Socket。

(6) 继续监听下一次的 Socket 连接。

当 ServerSocket 使用完毕后，应使用 ServerSocket 的 close()方法关闭该 ServerSocket。

ServerSocket 类提供的几个构造方法说明：

- ServerSocket(int port)方法：用指定端口(port)来创建一个 ServerSocket，该端口应该是一个有效端口的整数值(0～65535)。为了避免与其他程序产生冲突，通常绑定 1024 以上的端口。
- ServerSocket(int port,int backlog)方法：增加了一个用来改变队列长度的 backlog 参数。
- SetverSocket(int port,int backlog,InetAddress localAddr)：在机器存在多个 IP 的情况下，允许通过 localAddr 这个参数将 ServerSocket 绑定到指定的这个 IP 地址。

8.1.3 案例——简单聊天室

本例使用 Socket 通信实现一个简单聊天室。Socket 通信需要分别实现客户端、服务器端。

新建一个 Java 项目 ch8_1Server，在 src 下创建包 com. qsd，在包 com. qsd 下新建一个 class 文件 Server. java。

新建一个 Android 应用项目 ch8_1Client，包名为 com. qsd。在包 com. qsd 下新建一个 class 文件 ClientActivity. java。

1. 服务器端代码(Server. java)

```
package com.qsd;
import java.io.BufferedReader;
import java.io.InputStream;
import java.io.InputStreamReader;
import java.io.OutputStream;
import java.io.PrintWriter;
import java.net.ServerSocket;
import java.net.Socket;
public class Server {
    private int ServerPort = 9780;                          // 指定通信端口
    private ServerSocket serversocket = null;
    private OutputStream outputStream = null;
```

```
    private InputStream inputStream = null;
    private PrintWriter printWriter = null;
    private Socket socket = null;
    private BufferedReader bufferedreader = null;
    // Server 类的构造函数
    public Server() {
        try {
            // 根据指定的端口号,创建套接字
serversocket = new ServerSocket(ServerPort);
            System.out.println("服务器端准备就绪!");
            socket = serversocket.accept();        // 用 accept()方法等待客户端的连接
            System.out.println("客户端连接成功,可以通信\n");
        } catch (Exception exp) {
            exp.printStackTrace();                 // 异常处理
        }
        try {
            // 获取套接字输出流、输入流
            outputStream = socket.getOutputStream();
            inputStream = socket.getInputStream();
            printWriter = new PrintWriter(outputStream, true);
            bufferedreader = new BufferedReader(new InputStreamReader(inputStream));
            BufferedReader in = new BufferedReader(new InputStreamReader(System.in));
Date date = new Date();
            SimpleDateFormat df = new SimpleDateFormat("yyyy - MM - dd HH:mm");
            //设置日期格式
            String time = df.format(date);
            while (true) {
                // 接收客户端信息
                String message = bufferedreader.readLine();
                // 输出信息
                System.out.println(df.format(new Date()) + "客户端: " + message);
                // 结束聊天
                if (message.equals("88"))
                    break;
                message = in.readLine();
                printWriter.println(message);
            }
            outputStream.close();
            inputStream.close();
            socket.close();                        // 关闭套接字
            serversocket.close();                  // 关闭服务器套接字
            System.out.println("客户端已离开");
        } catch (Exception exp) {
            exp.printStackTrace();                 //异常处理
        } finally{}
    }
    //服务器端程序入口
    public static void main(String[] args) {
        new Server();
    }
}
```

2. 客户端代码(ClientActivity.java)

```
package com.qsd;
import java.io.BufferedReader;
import java.io.BufferedWriter;
import java.io.InputStreamReader;
import java.io.OutputStreamWriter;
import java.io.PrintWriter;
import java.net.InetAddress;
import java.net.Socket;
import com.qsd.R;
import android.app.Activity;
import android.app.AlertDialog;
import android.content.DialogInterface;
import android.os.Bundle;
import android.os.Handler;
import android.os.Message;
import android.util.Log;
import android.view.View;
import android.widget.Button;
import android.widget.EditText;
import android.widget.TextView;
public class ClientActivity extends Activity implements Runnable {
    //声明组件变量
    private TextView chatmsg = null;
    private EditText sendmsg = null;
    private Button sendbtn = null;
    private static final String HOST = "219.218.19.166";
    //服务器端IP地址也可以用10.0.2.2,它是模拟器设置的特定IP地址
    private static final int PORT = 4599;            //服务器端口号
    private Socket socket = null;
    private BufferedReader bufferedReader = null;
    private PrintWriter printWriter = null;
    private String string = "";
    public void onCreate(Bundle savedInstanceState) {
        super.onCreate(savedInstanceState);
        setContentView(R.layout.activity_main);
        chatmsg = (TextView) this.findViewById(R.id.chatmsg);
        sendmsg = (EditText) this.findViewById(R.id.sendmsg);
        sendbtn = (Button) this.findViewById(R.id.sendbtn);
        try {
            //指定IP和端口号,创建套接字
            socket = new Socket(HOST, PORT);
bufferedReader = new BufferedReader(new InputStreamReader(socket.getInputStream()));
printWriter = new PrintWriter(new BufferedWriter(new
```

```
OutputStreamWriter(socket.getOutputStream())), true);
        } catch (Exception e) {
            e.printStackTrace();                    // 异常处理
            CreateDialog(e.getMessage());           // 调用 CreateDialog()方法生成对话框
        }
        // 注册 sendbtn 的单击监听器
        sendbtn.setOnClickListener(new Button.OnClickListener() {
            public void onClick(View view) {
                String message = sendmsg.getText().toString();
        // 判断 socket 是否连接
                if (socket.isConnected()) {
                    if (!socket.isOutputShutdown()) {
                        printWriter.println(message);
Date date = new Date();
                SimpleDateFormat df = new SimpleDateFormat("yyyy - MM - dd HH:mm");
chatmsg.setText(chatmsg.getText().toString() + "\n" + df.format(new Date())
    + "Client: " + message);
                        sendmsg.setText("");
                    }
                }
            }
        });
// 启动线程
        new Thread(this).start();
    }
    // CreateDialog 产生对话框
    public void CreateDialog(String msmessage) {
        android.app.AlertDialog.Builder builder = new
            AlertDialog.Builder(this);
        builder.setTitle("出现异常");
        builder.setMessage(msmessage);
        builder.setPositiveButton("Yes", new DialogInterface.OnClickListener() {
            public void onClick(DialogInterface dialog, int which) {
            }
        });
        builder.setNegativeButton("No", new DialogInterface.OnClickListener() {
            public void onClick(DialogInterface dialog, int which) {
            }
        });
        builder.show();                                 // 显示对话框
    }//调用 run()方法运行线程
    public void run() {
try {
            while (true) {
                    if (socket.isConnected()) {
                    if (!socket.isInputShutdown()) {
                        if ((string = bufferedReader.readLine()) != null) {
```

```
                        Log.i("TAG", "++" + string);
                        string += " ";
                        messager.sendMessage(messager.obtainMessage());
                    } else {}
                }
            }
        }
    } catch (Exception exp) {
        exp.printStackTrace();                  // 显示异常信息
        Log.w("TAG", "-- " + exp.toString());
    }
}
//创建 Handler 对象 messager
    public Handler messager = new Handler() {
        public void handleMessage(Message msg) {
            super.handleMessage(msg);
            Log.i("TAG", "-- " + msg);           // 显示异常信息
Date date = new Date();
            SimpleDateFormat df = new SimpleDateFormat("yyyy - MM - dd HH:mm");
            chatmsg.setText(chatmsg.getText().toString() + "\n" +
df.format(new Date()) + "Server: " + string);
}
};
}
```

3. 布局文件(activity_main.xml)

```
<?xml version = "1.0" encoding = "utf - 8"?>
<LinearLayout xmlns:android = "http://schemas.android.com/apk/res/android"
    android:orientation = "vertical" android:layout_width = "fill_parent"
    android:layout_height = "fill_parent">
    <TextView android:id = "@ + id/chatmsg" android:layout_width = "fill_parent"
        android:layout_height = "wrap_content" android:text = "聊天室" />
    <EditText android:text = "" android:id = "@ + id/sendmsg"
        android:layout_width = "200dp" android:layout_height =
            "wrap_content"></EditText>
    <Button android:text = "发送" android:id = "@ + id/sendbtn"
        android:layout_width = "100dp"
            android:layout_height = "wrap_content"></Button>
</LinearLayout>
```

4. 在清单文件中添加权限

要让客户端能够访问服务器，必须在清单文件 AndroidManifest.xml 中添加权限：

```
<uses - permission android:name = "android.permission.INTERNET">
  </uses - permission>
```

5. 运行程序

在服务器端：Run As / Java Application
在客户端：Run As / Android Application
客户端的运行结果如图 8-2 所示，服务器端的输出结果如图 8-3 所示。

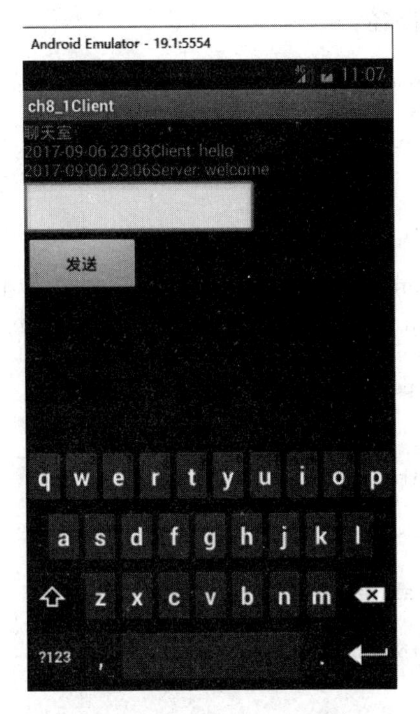

图 8-2　客户端的显示结果

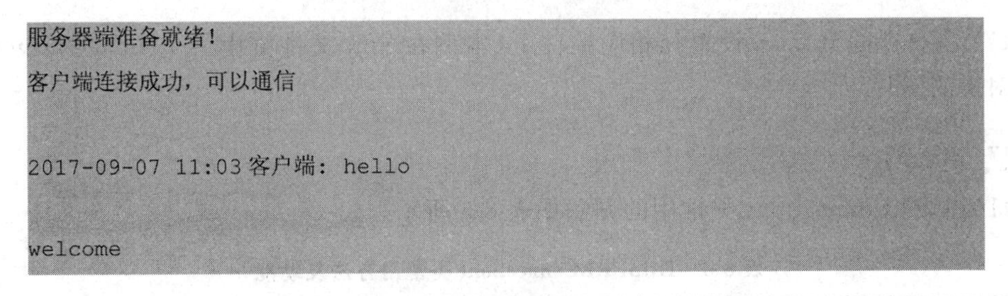

图 8-3　服务器端输出结果

8.2　基于 HTTP 的网络编程

在 Internet 上，基于 HTTP(Hyper Text Transfer Protocol，超文本传输协议)的应用是最为广泛的。同样，在移动互联时代，HTTP 将继续发挥它的重要作用。

在 Android 中，针对 HTTP 进行网络编程主要有以下两种方式：

- HttpURLConnection；
- Apache HTTP 开源客户端组件 HttpClient。

8.2.1　HttpURLConnection 的使用方法

若用户知道网络上某个资源的 URL，就可以直接使用 URL 来进行网络连接。使用 HttpURLConnection 访问网络的步骤如下。

(1) 设置要读取的资源路径：

```
String urlStr = " http://www.qrnu.edu.cn/themes/18118/
default/fullJQuery/img/06.jpg ";
```

(2) 实例化 URL：

```
URL url  =  new URL(urlStr);
```

(3) 获得 URLConnection 连接。由于 URLConnection 为抽象类，其对象不能直接实例化，一般通过 openConnection 方法获得：

```
URLConnection conn = url.openConnection();
conn.connect();
```

(4) 取得返回的 InputStream：

```
InputStream inputstream  = conn.getInputStream();
```

(5) 将获取的 InputStream 传送给 Bitmap：

```
Bitmap bitmap  = BitmapFactory.decodeStream(inputstream);
```

(6) 关闭流操作：

```
inputstream.close();
```

(7) 将获得的 Bitmap 设置到相应控件上(本例在 xml 文件中由一个 ImageView 控件显示对应的图片)：

```
myImageView.setImageBitmap(bitmap);
```

HttpURLConnectiont 类常用的方法如表 8-3 所示。

表 8-3　HttpURLConnectiont 类常用方法及功能

方　法	功能描述
InputStream getInputStream()	返回由此打开的连接读取的输入流
OutputStream getOutputStream()	返回写到此连接的输出流
String getRequestMethod()	获取请求方法
int getResponseCode()	获取状态码
void setRequestMethod(String method)	设置 URL 请求的方法
void setDoInput(boolean doinput)	设置输入流
void setDoOutput(boolean dooutput)	设置输出流
void setUseCaches(boolean usecaches)	设置连接是否使用任何可用的缓存
void disconnect()	关闭连接

8.2.2 案例——网络图片浏览器(使用 HttpURLConnectiont)

大多数 Android 网络应用，都需要与服务器进行通信。本例通过“网络图片浏览器”向读者展示手机端与服务器进行通信的过程。Android 的媒体库已经完成了图片、音视频解码。图片解码主要是由 BitmapFactory 库来完成，解码处理的位图由 imageView 控件展示。

新建一个 Android 应用项目：PhotoBrowser。

1. 布局文件(activity_main.xml)

设计"网络图片浏览器"需要创建用户交互界面，其对应的布局文件(activity_main.xml)如下：

```
<LinearLayout xmlns:android = "http://schemas.android.com/apk/res/android"
    xmlns:tools = "http://schemas.android.com/tools"
    android:layout_width = "match_parent"
    android:layout_height = "match_parent"
    android:orientation = "vertical"
    tools:context = ".MainActivity" >
    <ImageView
        android:id = "@ + id/imageview"
        android:layout_weight = "1100"
        android:layout_width = "fill_parent"
        android:layout_height = "fill_parent"
         />
    <EditText
        android:id = "@ + id/edittext"
        android:layout_width = "fill_parent"
        android:layout_height = "wrap_content"
        android:text = " http://www.qrnu.edu.cn/images/tp/06.jpg "
        android:singleLine = "true" />
    <Button
        android:layout_width = "fill_parent"
        android:layout_height = "wrap_content"
        android:onClick = "click"
        android:text = "打开图片" />
</LinearLayout>
```

2. 界面交互代码(MainActivity.java，包名 com.qsd)

```
package com.qsd;
import java.io.InputStream;
import java.net.HttpURLConnection;
import java.net.URL;
import com.qsd.R;
import android.app.Activity;
import android.graphics.Bitmap;
import android.graphics.BitmapFactory;
import android.os.Bundle;
import android.os.Handler;
import android.os.Message;
import android.text.TextUtils;
import android.view.View;
import android.widget.EditText;
```

```
import android.widget.ImageView;
import android.widget.Toast;
public class MainActivity extends Activity {
protected static final int CHANGE_UI = 2;
    protected static final int ERROR = 3;
    private EditText edittext;
    private ImageView imageview;
private Bitmap bitmap;
// 主线程创建消息处理器
    private Handler handler = new Handler(){
        public void handleMessage(android.os.Message msg) {
            if(msg.what == CHANGE_UI){
                imageview.setImageBitmap(bitmap);
            }else if(msg.what == ERROR){
                Toast.makeText(MainActivity.this, "显示图片错误", 0).show();
            }
        };
    };
    protected void onCreate(Bundle savedInstanceState) {
        super.onCreate(savedInstanceState);
        setContentView(R.layout.activity_main);
        edittext = (EditText) findViewById(R.id.edittext);
        imageview = (ImageView) findViewById(R.id.imageview);
    }
    public void click(View view) {
        final String path = edittext.getText().toString().trim();
        if (TextUtils.isEmpty(path)) {
            Toast.makeText(this, "图片路径为空,无法提取图片", 0).show();
        } else {
                    new Thread() {
                public void run() {
                    // Http 协议 get 方式从服务器获取图片
                    try {
                        URL url = new URL(path); //创建 URL 对象
                        // 打开 URL 对应的资源输入流
                        InputStream is = url.openStream();
                        //从 InputStream 中解析出图片
                        bitmap = BitmapFactory.decodeStream(is);
                        //发送消息,通知 UI 组件显示该图片
                        handler.sendEmptyMessage(CHANGE_UI);
                        is.close();
                    } catch (Exception e) {
                        e.printStackTrace();
                        Message message = new Message();
                        message.what = ERROR;
                        handler.sendMessage(message);
                    }
                };
            }.start();
        }
    }
}
```

3. 在清单文件添加权限

因为网络图片浏览器需要联网，所以要在清单文件 AndroidManifest. xml 中添加权限：

```
<uses-permission android:name="android.permission.INTERNET">
  </uses-permission>
```

4. 运行程序浏览图片

在文本框内输入图片的网络地址，单击“打开图片”按钮，打开该网络图片，如图 8-4 所示。

从图 8-4 可以看出，使用 HttpURLConnection 的 GET 方式请求指定图片地址，就能从网络服务器打开指定的图片。

图 8-4 浏览网络图片

运行方式：Run As/Android Application。

8.2.3 HttpClient 的使用方法

HttpClient 是 Apache Jakarta Common 下的子项目，它是 HTTP 客户端组件，对 Java. net 中的类进行封装和抽象，更适合在 Android 上开发网络应用，使得针对 HTTP 编程更加方便、高效。

使用 HttpClient 访问网络的步骤如下：

(1) 创建 HttpClient 对象；

(2) 指定访问网络的方式，创建一个 HttpGet 对象或者 HttpPost 对象；

(3) 如果需要发送请求参数，可调用 HttpGet、HttpPost 的 setParams()方法；

(4) 调用 HttpClient 对象的 execute()方法访问网络；

(5) 调用 HttpResponse. getEntity()方法获取 HttpEntity 对象。

HttpClient 的常用类如表 8-4 所示。

表 8-4 HttpClient 常用类

常用类名称	功能描述
HttpClient	请求网络的接口
DefaultHttpClient	实现了 HttpClient 接口的类
HttpGet	使用 GET 方式请求时,需创建该类实例
HttpPost	使用 POST 方式请求时,需创建该类实例
NameValuePair	关联参数的 Key、Value
BasicNameValuePair	以 Key、Value 的形式存放参数的类
UrlEncodeFormEntity	对提交给服务器的参数进行编码的类
HttpResponse	封装了服务器返回信息的类(包含头信息)
HttpEntity	封装了服务返回数据的类

8.2.4 案例——网络图片浏览器(使用 HttpClient)

为了让读者更好地掌握 HttpClient 的用法,本例改写 8.2.2 节中的案例"网络图片浏览器",用 HttpClient 实现同样的功能。

1. 布局文件(activity_main.xml)

布局文件同 8.2.2 节中的布局文件基本一致,只需修改图片地址为:

```
http://www.qrnu.edu.cn/images/tp/01.jpg
```

2. 界面交互代码(MainActivity.java)

```
package com.qsd;
import java.io.InputStream;
import org.apache.http.HttpEntity;
import org.apache.http.HttpResponse;
import org.apache.http.client.HttpClient;
import org.apache.http.client.methods.HttpGet;
import org.apache.http.impl.client.DefaultHttpClient;
import com.qsd.R;
import android.app.Activity;
import android.graphics.Bitmap;
import android.graphics.BitmapFactory;
import android.os.Bundle;
import android.os.Handler;
import android.os.Message;
import android.text.TextUtils;
import android.view.View;
import android.widget.EditText;
import android.widget.ImageView;
import android.widget.Toast;
public class MainActivity extends Activity {
    protected static final int T = 1;
    protected static final int F = 0;
    protected static final int S = 200;
```

```
    private EditText edittext;
    private ImageView imageview;
    private Handler handler = new Handler(){        //主线程创建消息处理器
        public void handleMessage(android.os.Message message) {
            if(message.what == T){
                Bitmap bitmap = (Bitmap) message.obj;
                imageview.setImageBitmap(bitmap);
            }else if(message.what == F){
                Toast.makeText(MainActivity.this, "显示异常", 0).show();
            }
        };
    };
protected void onCreate(Bundle savedInstanceState) {
        super.onCreate(savedInstanceState);
        setContentView(R.layout.activity_main);
        edittext = (EditText) findViewById(R.id.edittext);
        imageview = (ImageView) findViewById(R.id.imageview);
}public void click(View view) {
        final String path = edittext.getText().toString().trim();
        if (TextUtils.isEmpty(path)) {
            Toast.makeText(this, "图片所在路径不能为空", 0).show();
        } else {
            new Thread(new Runnable() {
                public void run() {

          HttpClient client = new DefaultHttpClient();        //创建 HttpClient 对象
        HttpGet httpGet = new HttpGet(path); //用 get 方式请求网络
        try {
            //获取返回的 HttpResponse 对象
            HttpResponse httpResponse = client.execute(httpGet);
            //检验服务器返回的状态码是否为 200
             if(httpResponse.getStatusLine().getStatusCode() == S){
             HttpEntity entity = httpResponse.getEntity();       //获取 HttpEntity 对象
              //获取输入流,获取 bitmap 对象
              InputStream content = entity.getContent();
              Bitmap bitmap = BitmapFactory.decodeStream(content);
                Message message = new Message();       //通知主线程更改 UI 界面
                message.what = T;
                message.obj = bitmap;
                handler.sendMessage(message);
            }
        } catch (Exception e) {
            e.printStackTrace();
Message message = new Message();
            message.what = F;
            handler.sendMessage(message);
        }
    }
}).start();
}
        }
    }
}
```

3. 在清单文件添加权限

因为网络图片浏览器需要联网，所以要在清单文件 AndroidManifest.xml 中添加权限：

```
<uses-permission android:name="android.permission.INTERNET">
  </uses-permission>
```

4. 运行程序浏览图片

在文本框内输入图片的网络地址，单击“打开图片”按钮，打开该网络图片，如图 8-5 所示。

图 8-5 浏览网络图片

从图 8-5 可以看出，使用 HttpClient 的 GET 方式请求指定图片地址，就能从网络服务器打开指定的图片。从实现过程可以发现，使用 HttpClient 提取网络数据更简洁、更高效。

运行方式：Run As/Android Application。

8.3 基于 WebView 的网络编程

8.3.1 WebView 视图组件

Android 提供了使用开源 WebKit 引擎的浏览器。Android 平台的 WebKit 由 Java 层和 WebKit 库两部分组成，Java 层负责与 Android 应用层进行通信，WebKit 类库则负责网页排版处理。

WebView 是 WebKit 中专门用来浏览网页的视图组件，用来显示网页或者显示应用的在线内容。它为用户提供了一系列的网页浏览、用户交互接口，通过这些接口显示和处理请求的网络资源。WebView 视图组件提供了一些比较常用的浏览方法，例如 loadUrl()和 loadData()。

使用 Android 提供的 WebView 组件可以将 HTML 内容转化为 Spanned 格式在

TextView 中进行显示。但如果应用要显示的内容只是一部分 HTML 片段，利用 TextView 进行显示效率更高。

WebView 提供了很多方法，例如，可以使用 loadUrl(String url)方法加载所要打开的网页等，如表 8-5 所示。

表 8-5 WebView 常用的方法及功能

方 法	功能描述
loadUrl()	按指定的 URL 打开一个 Web 页面资源
loadData()	加载 HTML 格式的网页内容
getSettings()	提取 WebView 的设置对象
addJavascriptInterface()	将一个对象添加到 JavaScript 的全局对象 window 中
clearCache()	清除缓存
destory()	销毁 WebView 组件

8.3.2 案例——使用 WebView 浏览网页

1. 布局文件(activity_main.xml)

```
<RelativeLayout xmlns:android = "http://schemas.android.com/apk/res/android"
    xmlns:tools = "http://schemas.android.com/tools"
    android:layout_width = "match_parent"
    android:layout_height = "match_parent"
    android:paddingBottom = "@dimen/activity_vertical_margin"
    android:paddingLeft = "@dimen/activity_horizontal_margin"
    android:paddingRight = "@dimen/activity_horizontal_margin"
    android:paddingTop = "@dimen/activity_vertical_margin"
    tools:context = "com.qsd.MainActivity" >
     <WebView android:id = "@ + id/show"
       android:layout_width = "match_parent"
       android:layout_height = "match_parent"/>
</RelativeLayout>
```

2. 打开网页代码(MainActivity.java)

```
package com.qsd;
iimport android.app.Activity;
import android.os.Bundle;
import android.view.KeyEvent;
import android.webkit.WebView;
import android.widget.EditText;
public class MainActivity extends Activity {
    WebView show;
    @Override
    protected void onCreate(Bundle savedInstanceState) {
        super.onCreate(savedInstanceState);
        setContentView(R.layout.activity_main);
```

```
            // 获取页面中 WebView 的组件
             show = (WebView)findViewById(R.id.show);
            // 加载并显示 url 对应的网页
             String url = "https://hao.360.cn";
                    show.loadUrl(url);
        }
}
```

3. 在清单文件添加权限

因为打开浏览器需要联网，所以要在清单文件 AndroidManifest.xml 中添加权限：

```
<uses-permission android:name="android.permission.INTERNET">
  </uses-permission>
```

4. 运行程序浏览 360 导航页面（见图 8-6）

运行方式：Run As / Android Application。

图 8-6 使用 WebView 浏览 360 导航页面

8.4 本章小结

本章详细讲解了 Android 操作系统的网络通信技术。首先讲解了基于 TCP/IP 的 Socket（称作“套接字”）通信技术，并实现了一个简单的聊天室项目。随后介绍了基于 HTTP 的两种访问网络方式：HttpURLConnection 和 HttpClient，并分别实现了网络图片浏览的项目。最后介绍了 WebView 组件，它是 WebKit 中专门用来浏览网页的视图组件。

8.5 练习题

一、填空题

1. 为了根据下载进度实时更新 UI 界面,需要用到 Handler 消息机制来实现________。
2. 与服务器交互过程中,最常用的两种数据提交方式是________和________。
3. 在 Android 中,针对 HTTP 进行网络通信的方法有________和________。

二、选择题

1. 下列通信方式中,不是 Android 系统提供的是(　　)。
   A. HTTP 通信　　B. Socket 通信
   C. URL 通信　　D. 以太网通信
2. 关于 GET 和 POST 请求方式,下列选项描述不正确的是(　　)。
   A. HTTP 规定 GET 方式请求 URL 的长度不能超过 1 KB
   B. 使用 POST 方式访问网络 URL 是有长度限制的
   C. 使用 GET 方式访问网络 URL 是有长度限制的
   D. GET 请求方式向服务器提交的参数跟在请求 URL 之后

三、简答题

1. 简述使用 Socket 访问网络的步骤。
2. 简述使用 HttpURLConnection 访问网络的步骤。
3. 简述使用 HttpClient 访问网络的步骤。
4. 简单说明 Android 开源 Web 浏览器引擎 WebKit 技术。

第9章 移动办公软件系统

本章重点

- 掌握 Android 移动办公软件系统项目架构
- 掌握移动办公软件系统登录页面的设计,包括首页下半部分包含的通知公告模块、工作日志模块、考勤管理模块、费用申请模块、请假模块和设置模块这六大模块;首页上半部分包含的日期时间、定位、天气这三大功能
- 掌握通知公告模块的设计,包括公告列表、功能详情
- 掌握工作日志模块的设计,包括工作内容、图片选择、定位
- 掌握考勤管理模块的设计,包括定位(每天只能提交一次)
- 掌握费用申请模块的设计,包括费用审批列表、费用申请
- 掌握请假模块的设计,包括请假列表、请假申请
- 掌握设置模块的设计,包括修改密码、用户退出

9.1 项目架构

9.1.1 项目架构

开发一个 APP 和盖房子差不多,第一件事就是为房子起个名字,搭一个框架,之后再往里面添砖加瓦。用 Eclipse 新建一个 Android 项目,为我们的移动办公考勤系统起一个项目名称,叫 MobileOffice,如图 9-1 所示。项目的名字起完了,下一步要给房子搭框架,框架的格式如图 9-2 所示,包名为 com. qsd. kqxt,其他使用系统默认即可。

整个项目放在 com. qsd. kqxt 下,其余部分包的说明如下:

- bean:所有的实体类(一般以 xxxInfo. java 结尾)放到这个包下。
- http:网络请求的一些类放到这个包下,主要用于网络请求公共方法。
- ui:这个主要是放自定义控件、弹出框等非关联 activity 的窗体。
- utils:所有的公用方法都放在这个包下。

9.1.2 其他命名规则

Android 资源文件夹 res 中的各个文件也有不同的命名要求,例如背景图片资源一般是 bg_xxx. png、图标一般是 btn_xxx. png 等。一般的命名规则如表 9-1 所示。

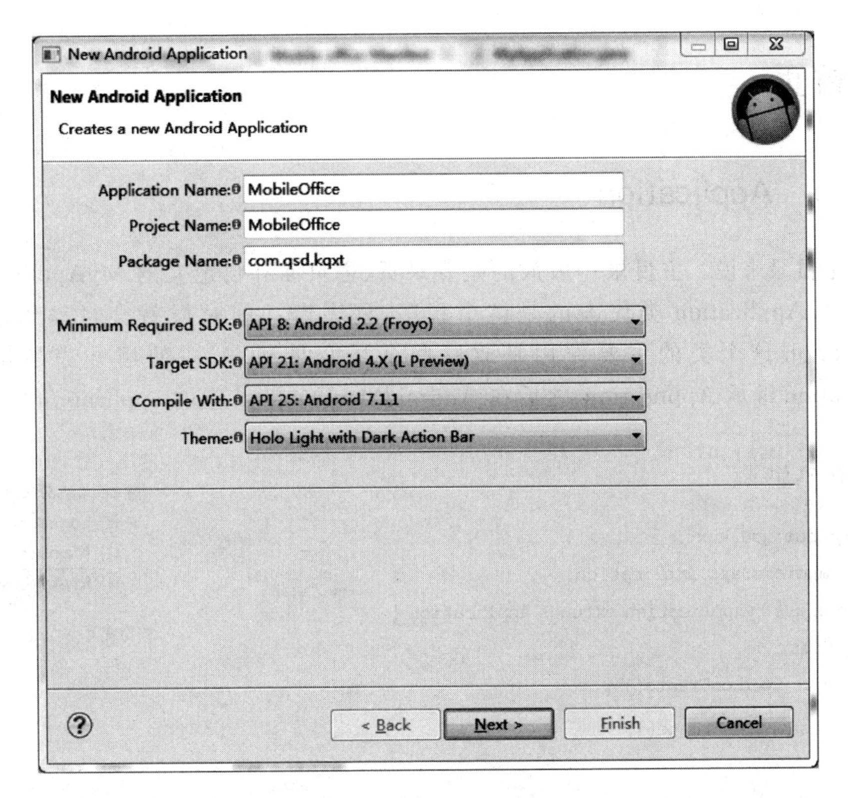

图 9-1

src
- com.qsd.kqxt
- com.qsd.kqxt.bean
- com.qsd.kqxt.common
- com.qsd.kqxt.ctl
- com.qsd.kqxt.Fragment
- com.qsd.kqxt.http
- com.qsd.kqxt.ui
- com.qsd.kqxt.utils
- com.qsd.kqxt.weight

图 9-2

表 9-1　其他命名规则

命名规则	目　录	说　明
bg_xxx. png	res/drawable	背景图资源
ic_xxx. png	res/drawable	显示型图标资源
btn_xxx. png	res/drawable	按钮图标资源
activity_xxx. xml	res/layout	对应 xxxActivity. java 类的文件
fragment_xxx. xml	res/layout	对应 xxxFragment. java 类的文件
Item_xxx. xml	res/layout	对应实体类列表的列表项文件
view_xxx. xml	res/layout	对应自定义控件的文件

9.2 首页

9.2.1 Application

先从主目录来说。主目录一般放的是登录页、主页面和自定义的 MyApplication。如图 9-3 所示，Application 类代表的是整个程序，程序启动的时候就会自动创建。其中 MyApplication 这个类的主要作用是放一些全局变量和方法，使用方法是新建一个 MyApplication 继承 Application，之后在 AndroidManifes. xml 中的 application 的 name 中声明。

示例代码如下：

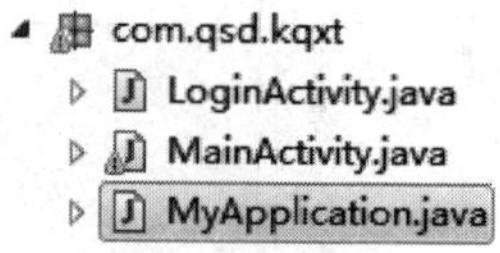

com.qsd.kqxt
LoginActivity.java
MainActivity.java
MyApplication.java

图 9-3 主目录

```
package com.qsd.kqxt;
import android.app.Application;
public class MyApplication extends Application {
    @Override
    public void onCreate() {
        super.onCreate();
    }
}
```

修改 AndroidManifes. xml 中的代码如下：

```
<application
        android:name = "com.qsd.kqxt.MyApplication"
        android:icon = "@drawable/ic_launcher"
android:label = "@string/app_name">
</application>
```

9.2.2 LoginActivity(登录页面)

先看一下登录页面的效果，如图 9-4 所示。效果图中布局比较简单，主要可以分为两块内容，一块是上部分的图片，一块是下部分的输入框等控件。

把图片放到 res/drawable-hdpi 目录下，其中用到的资源图片如表 9-2 所示。

在 res/layout 目录下新建一个布局文件 activity_login. xml。在编写代码之前一般先要想好一个预设的布局轮廓，哪个布局先写。activity_login. xml 的布局轮廓如图 9-5 所示。

activity_login. xml 中用到了 CheckBox 的自定义控件属性。CheckBox 自定义属性的使用方法是，先在 res 目录下新建文件夹 drawable，之后在 drawable 目录下新建 radiobtn_selector. xml，在 Root Element 中选择 selector，最后确定。

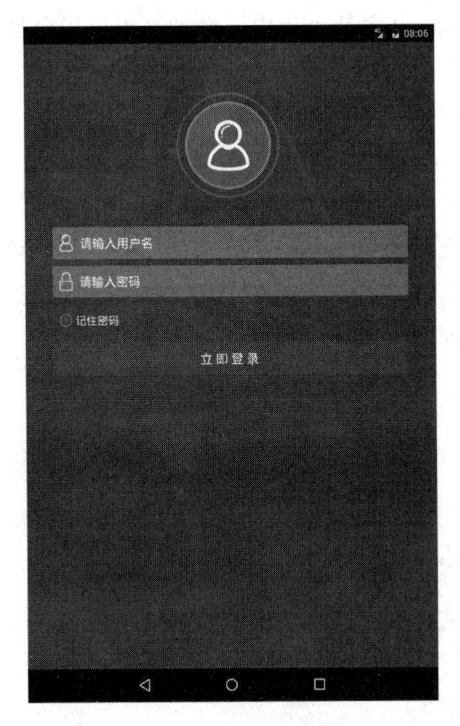

图 9-4　登录页面布局

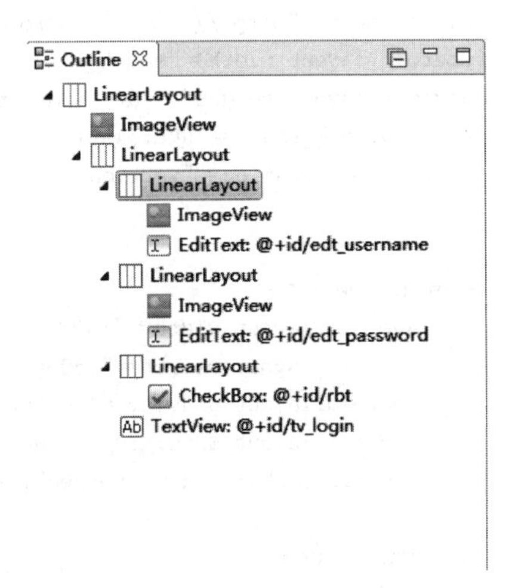

图 9-5　布局的轮廓

表 9-2　图片资源

图片名称	图　片	说　明
bg_login		登录页的背景图片
bg_touxiang		头像，白色图标(为了在 Word 下易于辨识填充了底色)
radiobg_normal		单选按钮，未选中状态
radiobg_press		单选按钮，选中状态
ic_user		用户，白色图标(为了在 Word 下易于辨识填充了底色)
ic_pass		密码，白色图标(为了在 Word 下易于辨识填充了底色)

radiobtn_selector.xml 代码如下：

```
<?xml version = "1.0" encoding = "utf - 8"?>
<selector xmlns:android = "http://schemas.android.com/apk/res/android">

    <item android:drawable = "@drawable/radiobg_press"
        android:state_checked = "true" android:state_enabled = "true"></item>
    <item android:drawable = "@drawable/radiobg_normal"
        android:state_checked = "false" android:state_enabled = "true"></item>
    <item android:drawable = "@drawable/radiobg_normal"/>
</selector>
```

在 selector 节点中，item 用于添加图片资源，包含是否选中属性和是否可用属性，最后一个 item 只添加图片资源，其他属性选择默认状态。之后，在 CheckBox 中通过语句 android:button="@drawable/radiobtn_selector"予以使用。

activity_login.xml 主要代码如下：

```
<LinearLayout xmlns:android = "http://schemas.android.com/apk/res/android"
    xmlns:tools = "http://schemas.android.com/tools"
    android:layout_width = "match_parent"
    android:layout_height = "match_parent"
    android:background = "@drawable/bg_login"
    android:orientation = "vertical"
    >

    <ImageView
        android:layout_width = "150dp"
        android:layout_height = "150dp"
        android:layout_gravity = "center_horizontal"
        android:layout_marginTop = "70dp"
        android:background = "@drawable/bg_touxiang" />

    <LinearLayout
        android:layout_width = "match_parent"
        android:layout_height = "match_parent"
        android:layout_margin = "40dp"
        android:orientation = "vertical" >

        <LinearLayout
            android:layout_width = "match_parent"
            android:layout_height = "45dp"
            android:background = "@drawable/edt_bg"
            android:orientation = "horizontal"
            android:paddingLeft = "10dp" >

            <ImageView
                android:layout_width = "wrap_content"
                android:layout_height = "wrap_content"
                android:layout_gravity = "center_vertical"
                android:background = "@drawable/ic_user" />

            <EditText
                android:id = "@ + id/edt_username"
                android:layout_width = "match_parent"
                android:layout_height = "match_parent"
                android:background = "@null"
                android:gravity = "center_vertical"
                android:hint = "请输入用户名"
                android:paddingLeft = "10dp"
                android:textColor = "#ffffff"
                android:textColorHint = "#ffffff" />
        </LinearLayout>
```

```
<LinearLayout
    android:layout_width="match_parent"
    android:layout_height="45dp"
    android:layout_marginTop="10dp"
    android:background="@drawable/edt_bg"
    android:orientation="horizontal"
    android:paddingLeft="10dp" >

    <ImageView
        android:layout_width="wrap_content"
        android:layout_height="wrap_content"
        android:layout_gravity="center_vertical"
        android:background="@drawable/ic_pass" />

    <EditText
        android:id="@+id/edt_password"
        android:layout_width="match_parent"
        android:layout_height="match_parent"
        android:background="@null"
        android:gravity="center_vertical"
        android:hint="请输入密码"
        android:inputType="textPassword"
        android:paddingLeft="10dp"
        android:textColor="#ffffff"
        android:textColorHint="#ffffff" />
</LinearLayout>

<LinearLayout
    android:layout_width="match_parent"
    android:layout_height="45dp"
    android:layout_marginTop="10dp"
    android:gravity="center_vertical"
    android:orientation="horizontal"
    android:paddingLeft="10dp" >

    <CheckBox
        android:id="@+id/rbt"
        android:layout_width="wrap_content"
        android:layout_height="wrap_content"
        android:button="@drawable/radiobtn_selector"
        android:checked="true"
        android:paddingLeft="5dp"
        android:text="记住密码"
        android:textColor="#ffffff"
        android:textSize="16sp" />
</LinearLayout>

<TextView
    android:id="@+id/tv_login"
    android:layout_width="match_parent"
```

```
            android:layout_height = "45dp"
            android:layout_marginTop = "10dp"
            android:background = "@color/btn_submit"
            android:gravity = "center"
            android:text = "立 即 登 录"
            android:textColor = "#ffffff"
            android:textSize = "18sp" />
    </LinearLayout>

</LinearLayout>
```

LoginActivity.java 主要代码如下：

```
package com.qsd.kqxt;

import android.app.Activity;
import android.content.Intent;
import android.os.Bundle;
import android.util.Log;
import android.view.View;
import android.view.View.OnClickListener;
import android.widget.CheckBox;
import android.widget.EditText;
import android.widget.TextView;
import android.widget.Toast;

import com.qsd.kqxt.R;
import com.qsd.kqxt.utils.SPUtils;

public class LoginActivity extends Activity {
    private EditText musetnameEditText;
    private EditText mpasswordEditText;
    private CheckBox mRadioButton;
    private TextView mtv;

    @Override
    protected void onCreate(Bundle savedInstanceState) {
        super.onCreate(savedInstanceState);
        setContentView(R.layout.activity_login);
        initView();
    }

    private void initView() {
        musetnameEditText = (EditText) findViewById(R.id.edt_username);
        mpasswordEditText = (EditText) findViewById(R.id.edt_password);
        mRadioButton = (CheckBox) findViewById(R.id.rbt);
        mtv = (TextView) findViewById(R.id.tv_login);
        mtv.setOnClickListener(new OnClickListener() {

            @Override
            public void onClick(View v) {
```

```
                // TODO Auto - generated method stub
                switch (v.getId()) {
                case R.id.tv_login:
                    login();
                    break;
                }
            }
        });
    }

    private void login() {
        String name = musetnameEditText.getText().toString().trim();
        String pwd = mpasswordEditText.getText().toString().trim();
        if (name.equals("") || pwd.equals("")) {
            Toast.makeText(LoginActivity.this, "用户名或密码不能为空!",
              Toast.LENGTH_LONG).show();

            return;
        }
        if (name.equals("admin") && pwd.equals("123456")) {
            toMainActivity();
            saveData("admin");
        } else {
            Toast.makeText(LoginActivity.this, "用户名或密码錯誤!",
              Toast.LENGTH_LONG).show();
        }

    }

    private void toMainActivity() {
        Intent intent = new Intent(LoginActivity.this, MainActivity.class);
        LoginActivity.this.startActivity(intent);
        LoginActivity.this.finish();
    }

    private void saveData(String array) {
        Log.i("array", array.toString());
        if (mRadioButton.isChecked()) {
            SPUtils utils = new SPUtils(LoginActivity.this);
            utils.setData("adminID", array);
        }
    }

}
```

9.2.3 MainActivity(主页面)

登录之后是 MainActivity,此页面布局比较复杂。先看如图 9-6 所示的页面效果,拆分后的布局分为上下两个部分,上半部分主要是时间显示、定位以及当前的天气,下半部分是 6 个图文混排的按钮。

图 9-6　页面布局效果

把要用到的图片放到 res/drawable-hdpi 目录下。应用页面用到的图片资源如表 9-3 所示。

表 9-3　图片资源

图 片 名 称	图　　片	说　　明
bg_main		主页上半部分的背景图片
ic_btn1		通知公告图标
ic_btn2		工作日志图标
ic_btn3		考勤管理图标
ic_btn4		费用申请图标
ic_btn5		请假图标
ic_exit		设置图标
weather00		天气图标，此类图标有 33 个（为了在 Word 易于辨识填充了底色）

资源准备就绪后，编写在 res/layout 目录下的 activity_main.xml 文件，主要代码如下：

```
<LinearLayout xmlns:android = "http://schemas.android.com/apk/res/android"
    xmlns:tools = "http://schemas.android.com/tools"
```

```
android:layout_width = "match_parent"
android:layout_height = "match_parent"
android:orientation = "vertical"
tools:context = ".MainActivity" >

<LinearLayout
    android:layout_width = "match_parent"
    android:layout_height = "0dp"
    android:layout_weight = "1"
    android:background = "@drawable/bg_main"
    android:orientation = "vertical" >

    <TextView
        android:layout_width = "match_parent"
        android:layout_height = "wrap_content"
        android:layout_margin = "5dp"
        android:gravity = "center_horizontal"
        android:text = "移动办公管理系统 V1.0"
        android:textColor = "#ffffff"
        android:textSize = "14sp" />

    <LinearLayout
        android:layout_width = "match_parent"
        android:layout_height = "match_parent"
        android:orientation = "horizontal" >

        <LinearLayout
            android:layout_width = "0dp"
            android:layout_height = "match_parent"
            android:layout_marginLeft = "30dp"
            android:layout_weight = "1"
            android:gravity = "center_vertical"
            android:orientation = "vertical" >

            <TextView
                android:id = "@ + id/tv_time"
                android:layout_width = "wrap_content"
                android:layout_height = "wrap_content"
                android:text = "00:00"
                android:textColor = "#ffffff"
                android:textSize = "45sp" />

            <TextView
                android:id = "@ + id/tv_date"
                android:layout_width = "wrap_content"
                android:layout_height = "wrap_content"
                android:layout_marginTop = "10dp"
                android:text = "00.00 星期五·0000"
                android:textColor = "#ffffff"
                android:textSize = "18sp" />
        </LinearLayout>
```

```
<LinearLayout
    android:layout_width = "0dp"
    android:layout_height = "match_parent"
    android:layout_weight = "1"
    android:gravity = "center"
    android:orientation = "vertical" >

    <LinearLayout
        android:layout_width = "match_parent"
        android:layout_height = "wrap_content"
        android:gravity = "center"
        android:orientation = "horizontal" >

        <ImageView
            android:id = "@ + id/iv_weather"
            android:layout_width = "wrap_content"
            android:layout_height = "wrap_content"
            android:scaleType = "fitCenter"
            android:src = "@drawable/weather00" />

        <TextView
            android:id = "@ + id/tv_city"
            android:layout_width = "wrap_content"
            android:layout_height = "wrap_content"
            android:gravity = "center"
            android:paddingLeft = "5dp"
            android:paddingRight = "5dp"
            android:paddingTop = "10dp"
            android:textColor = "#fff"
            android:textSize = "16sp" />
    </LinearLayout>

    <TextView
        android:id = "@ + id/tv_weather"
        android:layout_width = "wrap_content"
        android:layout_height = "wrap_content"
        android:gravity = "center"
        android:text = "晴"
        android:textColor = "#fff"
        android:textSize = "16sp" />

    <TextView
        android:id = "@ + id/tv_temperature"
        android:layout_width = "match_parent"
        android:layout_height = "wrap_content"
        android:gravity = "center"
        android:paddingLeft = "5dp"
        android:paddingRight = "5dp"
        android:text = "24° - 29°"
        android:textColor = "#fff"
```

```
                android:textSize = "30sp" />
            </LinearLayout >
        </LinearLayout >
    </LinearLayout >

    < LinearLayout
        android:layout_width = "match_parent"
        android:layout_height = "0dp"
        android:layout_weight = "2"
        android:background = "@color/bg_all"
        android:orientation = "vertical" >

        < LinearLayout
            android:layout_width = "match_parent"
            android:layout_height = "wrap_content"
            android:layout_marginTop = "20dp"
            android:orientation = "horizontal" >

            < LinearLayout
                android:id = "@ + id/ll_main1"
                android:layout_width = "0dp"
                android:layout_height = "wrap_content"
                android:layout_weight = "1"
                android:gravity = "center_horizontal"
                android:orientation = "vertical" >

                < RelativeLayout
                    android:layout_width = "80dp"
                    android:layout_height = "80dp" >

                    < ImageView
                        android:layout_width = "80dp"
                        android:layout_height = "80dp"
                        android:layout_centerInParent = "true"
                        android:background = "@drawable/ic_btn1" />

                    < TextView
                        android:id = "@ + id/drop"
                        android:layout_width = "25dp"
                        android:layout_height = "25dp"
                        android:layout_alignParentRight = "true"
                        android:layout_alignParentTop = "true"
                        android:background = "@drawable/bg_read"
                        android:gravity = "center"
                        android:textColor = " # fff"
                        android:textSize = "10dp"
                        android:visibility = "invisible" />
                </RelativeLayout >

                < TextView
                    android:layout_width = "wrap_content"
```

```
                android:layout_height = "wrap_content"
                android:layout_marginTop = "5dp"
                android:text = "通知公告"
                android:textSize = "16sp" />
        </LinearLayout>

        <LinearLayout
            android:id = "@+id/ll_main2"
            android:layout_width = "0dp"
            android:layout_height = "wrap_content"
            android:layout_weight = "1"
            android:gravity = "center_horizontal"
            android:orientation = "vertical" >

            <ImageView
                android:layout_width = "80dp"
                android:layout_height = "80dp"
                android:background = "@drawable/ic_btn2" />

            <TextView
                android:layout_width = "wrap_content"
                android:layout_height = "wrap_content"
                android:layout_marginTop = "5dp"
                android:text = "工作日志"
                android:textSize = "16sp" />
        </LinearLayout>

        <LinearLayout
            android:id = "@+id/ll_main3"
            android:layout_width = "0dp"
            android:layout_height = "wrap_content"
            android:layout_weight = "1"
            android:gravity = "center_horizontal"
            android:orientation = "vertical" >

            <ImageView
                android:layout_width = "80dp"
                android:layout_height = "80dp"
                android:background = "@drawable/ic_btn3" />

            <TextView
                android:layout_width = "wrap_content"
                android:layout_height = "wrap_content"
                android:layout_marginTop = "5dp"
                android:text = "考勤管理"
                android:textSize = "16sp" />
        </LinearLayout>
    </LinearLayout>

    <LinearLayout
        android:layout_width = "match_parent"
```

```
android:layout_height = "wrap_content"
android:layout_marginTop = "20dp"
android:orientation = "horizontal" >

<LinearLayout
    android:id = "@ + id/ll_main4"
    android:layout_width = "0dp"
    android:layout_height = "wrap_content"
    android:layout_weight = "1"
    android:gravity = "center_horizontal"
    android:orientation = "vertical" >

    <ImageView
        android:layout_width = "80dp"
        android:layout_height = "80dp"
        android:background = "@drawable/ic_btn4" />

    <TextView
        android:layout_width = "wrap_content"
        android:layout_height = "wrap_content"
        android:layout_marginTop = "5dp"
        android:text = "费用申请"
        android:textSize = "16sp" />
</LinearLayout>

<LinearLayout
    android:id = "@ + id/ll_main5"
    android:layout_width = "0dp"
    android:layout_height = "wrap_content"
    android:layout_weight = "1"
    android:gravity = "center_horizontal"
    android:orientation = "vertical" >

    <ImageView
        android:layout_width = "80dp"
        android:layout_height = "80dp"
        android:background = "@drawable/ic_btn5" />

    <TextView
        android:layout_width = "wrap_content"
        android:layout_height = "wrap_content"
        android:layout_marginTop = "5dp"
        android:text = "请假"
        android:textSize = "16sp" />
</LinearLayout>

<LinearLayout
    android:id = "@ + id/ll_main6"
    android:layout_width = "0dp"
    android:layout_height = "wrap_content"
    android:layout_weight = "1"
```

```
                android:gravity = "center_horizontal"
                android:orientation = "vertical" >

                <ImageView
                    android:layout_width = "80dp"
                    android:layout_height = "80dp"
                    android:background = "@drawable/ic_exit" />

                <TextView
                    android:layout_width = "wrap_content"
                    android:layout_height = "wrap_content"
                    android:layout_marginTop = "5dp"
                    android:text = "设置"
                    android:textSize = "16sp" />
            </LinearLayout>
        </LinearLayout>
    </LinearLayout>

</LinearLayout>
```

编写 src/com.qsd.kqxt 下的 MainActivity.java 文件，主要代码如下：

```
package com.qsd.kqxt;

import java.util.Calendar;

import android.app.Activity;
import android.content.Intent;
import android.os.Bundle;
import android.os.Handler;
import android.view.View;
import android.view.View.OnClickListener;
import android.widget.LinearLayout;
import android.widget.TextView;

import com.qsd.kqxt.activity.CheckActivity;
import com.qsd.kqxt.activity.CostActivity;
import com.qsd.kqxt.activity.LeaveActivity;
import com.qsd.kqxt.activity.MessageActivity;
import com.qsd.kqxt.activity.SettingActivity;
import com.qsd.kqxt.activity.WorkActivity;
import com.qsd.kqxt.bean.LocationInfo;
import com.qsd.kqxt.bean.UserInfo;
import com.qsd.kqxt.utils.Location;
import com.qsd.kqxt.utils.SPUtils;

public class MainActivity extends Activity implements OnClickListener {

    public static Activity instance;

    SPUtils utils;
```

```
String city;
int hour;
boolean mState = true;
int count;      // 未读消息数
private TextView mTime;

private TextView mDate;

private TextView mCity;

private TextView mDrop;

private LinearLayout mMain1, mMain2, mMain3, mMain4, mMain5, mMain6;
Location location;
Handler handler = new Handler() {
    public void handleMessage(android.os.Message msg) {
        switch (msg.what) {
        case 0:
            LocationInfo info = (LocationInfo) msg.obj;
            city = info.getCity();
            if (city != null && !city.equals("")) {
                mCity.setText(city);
                if (mState) {
                    mState = false;
                }
            }
            break;
        case 1:
            initTime();
            handler.sendEmptyMessageDelayed(1, 5000);
            break;
        }
    };
};

@Override
protected void onCreate(Bundle savedInstanceState) {
    super.onCreate(savedInstanceState);
    setContentView(R.layout.activity_main);

    utils = new SPUtils(MainActivity.this);
    instance = this;
    checkLogin();
    handler.sendEmptyMessage(1);
    initView();
}

/***
 * 初始化控件
 */
private void initView() {
```

```
        mTime = (TextView) findViewById(R.id.tv_time);
        mDate = (TextView) findViewById(R.id.tv_date);
        mCity = (TextView) findViewById(R.id.tv_city);
        mDrop = (TextView) findViewById(R.id.drop);
        mMain1 = (LinearLayout) findViewById(R.id.ll_main1);
        mMain2 = (LinearLayout) findViewById(R.id.ll_main2);
        mMain3 = (LinearLayout) findViewById(R.id.ll_main3);
        mMain4 = (LinearLayout) findViewById(R.id.ll_main4);
        mMain5 = (LinearLayout) findViewById(R.id.ll_main5);
        mMain6 = (LinearLayout) findViewById(R.id.ll_main6);

        mMain1.setOnClickListener(this);
        mMain2.setOnClickListener(this);
        mMain3.setOnClickListener(this);
        mMain4.setOnClickListener(this);
        mMain5.setOnClickListener(this);
        mMain6.setOnClickListener(this);

        count = 2;
        if (count > 0) {
            // 未读信息大于 0 则在右上角显示数量,否则隐藏

            mDrop.setVisibility(View.VISIBLE);
            mDrop.setText(count + "");
        } else {
            mDrop.setVisibility(View.GONE);
        }
    }

    /***
     * 记住密码功能,即是否保存用户数据
     */
    private void checkLogin() {
        // 获取本地用户信息
        String adminId = utils.getData("adminID");
        // 如果没有数据则跳转到登录页面
        if ((adminId.equals("") || adminId == null)) {
            Intent intent = new Intent(MainActivity.this, LoginActivity.class);
            startActivity(intent);
            finish();
        } else {
            initData();
        }
    }

    @Override
    public void onClick(View v) {
        Intent intent = new Intent();
        switch (v.getId()) {
        case R.id.ll_main1:// 通知
```

```
            intent.setClass(this, MessageActivity.class);
            startActivity(intent);

            break;
        case R.id.ll_main2:// 工作日志
            intent.setClass(this, WorkActivity.class);
            startActivity(intent);
            break;
        case R.id.ll_main3:// 考勤管理
            intent.setClass(this, CheckActivity.class);
            startActivity(intent);
            break;
        case R.id.ll_main4:// 费用审批
            intent.setClass(this, CostActivity.class);
            startActivity(intent);
            break;
        case R.id.ll_main5:// 请假
            intent.setClass(this, LeaveActivity.class);
            startActivity(intent);
            break;
        case R.id.ll_main6:// 注销
            intent.setClass(MainActivity.this, SettingActivity.class);
            startActivity(intent);
            break;
        }

}

/***
 * 初始化数据
 */
private void initData() {
    location = new Location(handler, MainActivity.this);
    location.start();
}

/***
 * 初始化时间
 */
private void initTime() {
    Calendar c = Calendar.getInstance();
    int year = c.get(Calendar.YEAR);
    int month = c.get(Calendar.MONTH);
    month++;
    int day = c.get(Calendar.DAY_OF_MONTH);
    String dayString = day + "";
    if (day < 10) {
        dayString = "0" + day;
    }
    hour = c.get(Calendar.HOUR_OF_DAY);
    String hourString = hour + "";
```

```
        if (hour < 10) {
            hourString = "0" + hour;
        }
        int min = c.get(Calendar.MINUTE);
        String minString = min + "";
        if (min < 10) {
            minString = "0" + min;
        }
        int weekday = c.get(Calendar.DAY_OF_WEEK);
        String mWeekday = getWeek(weekday);
        mTime.setText(hourString + ":" + minString);
        mDate.setText(month + "." + dayString + " " + mWeekday + "·" + year);
    }

    private String getWeek(int week) {
        String mWeek = "";
        switch (week) {
        case Calendar.MONDAY:
            mWeek = "星期一";
            break;
        case Calendar.TUESDAY:
            mWeek = "星期二";
            break;
        case Calendar.WEDNESDAY:
            mWeek = "星期三";
            break;
        case Calendar.THURSDAY:
            mWeek = "星期四";
            break;
        case Calendar.FRIDAY:
            mWeek = "星期五";
            break;
        case Calendar.SATURDAY:
            mWeek = "星期六";
            break;
        case Calendar.SUNDAY:
            mWeek = "星期天";
            break;
        }
        return mWeek;
    }

}
```

这里用到的新的知识是百度地图定位，请读者自己去百度开发者中心(bsyun.baidu.com)上注册学习。这里封装了百度的定位功能，使用时只需要声明一下，再调用 start 方法就可以获得数据。在 utils 目录下新建一个类 Location.java，其主要代码如下：

```
package com.qsd.kqxt.utils;

import android.content.Context;
```

```
import android.os.Handler;
import android.os.Message;
import android.util.Log;

import com.baidu.location.BDLocation;
import com.baidu.location.BDLocationListener;
import com.baidu.location.LocationClient;
import com.baidu.location.LocationClientOption;
import com.baidu.location.LocationClientOption.LocationMode;
import com.qsd.kqxt.bean.LocationInfo;

public class Location {

    Handler handler;
    public LocationClient mLocationClient = null;
    public BDLocationListener myListener;
    LocationInfo info;
    Context context;

    public Location(Handler handler, Context context) {
        this.handler = handler;
        this.context = context;
        info = new LocationInfo();
        myListener = new MyLocationListener();
        //初始化 LocationClient 类
        mLocationClient = new LocationClient(context);      // 声明 LocationClient 类
        mLocationClient.registerLocationListener(myListener);      // 注册监听函数
        initLocation();
    }

    //开始定位
    public void start() {
        mLocationClient.start();
    }

    //停止定位
    public void stop() {
        mLocationClient.stop();
    }

    //配置定位 SDK 参数
    private void initLocation() {
        LocationClientOption option = new LocationClientOption();
        option.setLocationMode(LocationMode.Hight_Accuracy);
        // 设置定位模式,可选,默认为高精度,这里设置为高精度、低功耗、仅设备


        option.setCoorType("bd0911");
        // 设置返回的定位结果坐标系,可选,默认为 gcj02,这里设为 bd0911
        // gcj02 为火星坐标系,bd0911 为百度坐标系
```

```
        int span = 0;
        option.setScanSpan(span);
        //可选,默认为 0,即仅定位一次.
        //若 span = 1000,则设置发起定位请求的间隔需要大于等于 1000ms 有效
        option.setIsNeedAddress(true);
        // 设置是否需要地址信息,可选,默认不需要(false)

        option.setOpenGps(true);
        // 设置是否使用 GPS,可选,默认不使用(false)

        option.setLocationNotify(true);
        // 设置是否当 GPS 有效时按照 1 次每秒的频率输出 GPS 结果,可选,默认不输出(false)

        option.setIsNeedLocationDescribe(true);
        //设置是否需要位置语义化结果(结果类似于"在长春国贸大厦附近"),
        //可选,默认不需要(false),可以在 BDLocation.getLocationDescribe 里得到

        option.setIsNeedLocationPoiList(true);
        //设置是否需要 POI 结果,可选,默认不需要(false)
        //可以在 BDLocation.getPoiList 里得到

        option.setIgnoreKillProcess(false);
        // 设置是否在 stop 的时候杀死这个进程,可选,默认不杀死(true)
        //定位在 SDK 内部是一个 SERVICE,并被放到了独立进程中
        option.SetIgnoreCacheException(false);
        // 可选,默认 false,设置是否收集 CRASH 信息,默认收集

        option.setEnableSimulateGps(false);
        // 设置是否需要过滤 GPS 仿真结果,可选,默认需要(false)

        mLocationClient.setLocOption(option);
    }

    //实现 BDLocationListener 接口
    public class MyLocationListener implements BDLocationListener {

        @Override
        public void onReceiveLocation(BDLocation location) {
            String latitude = String.valueOf(location.getLatitude());
            String longitude = String.valueOf(location.getLongitude());
            String add = location.getAddrStr();
            String locDescribution = location.getLocationDescribe();
            info.setLatitude(String.valueOf(location.getLatitude()));
            info.setLongitude(String.valueOf(location.getLongitude()));
            info.setAdd(location.getAddrStr());
            info.setLocDescribution(location.getLocationDescribe());
            info.setCity(location.getCity());
            Log.i("location", "经度: " + longitude + "\n 纬度: " + latitude +
                    "\n 位置: " + location + locDescribution);
            Message msg = new Message();
            msg.what = 0;
```

```
            msg.obj = info;
            handler.sendMessage(msg);
        }

        @Override
        public void onConnectHotSpotMessage(String arg0, int arg1) {
            // TODO Auto - generated method stub

        }

    }

}
```

在 MainActivity.java 中还用到了一个封装好的类，名字叫 SPUtils，这个类也是放到 utils 目录下的，主要功能是保存用户信息。主要代码如下：

```
package com.qsd.kqxt.utils;

import android.content.Context;
import android.content.SharedPreferences;
import android.content.SharedPreferences.Editor;

public class SPUtils {

    SharedPreferences preferences;
    Editor editor;

    public SPUtils(Context context) {
        preferences = context
                .getSharedPreferences("MySP", Context.MODE_PRIVATE);
        editor = preferences.edit();
    }

    /***
     * 获取数据
     *
     * @param key
     *            键值
     * @return 数据
     */
    public String getData(String key) {
        String data = preferences.getString(key, "");
        return data;
    }

    /***
     * 保存数据
     *
     * @param key
     *            键值
     * @param value
     *        保存的数据
```

```
     */
    public void setData(String key, String value) {
        editor.putString(key, value);
        editor.commit();
    }

    /***
     * 清空所有数据
     */
    public void clearAllData() {
        editor.clear();
        editor.commit();
    }
}
```

9.3 通知公告模块

此模块的主要功能是发送通知信息和公告信息，信息内容主要是通知或公告的标题、内容及时间等。从设计效果图 9-7 中可以看到布局比较简单，包括一个头部和一个列表数据。在通知公告效果图 9-7 中单击某个标题内容，如单击标题内容 0 就会出现相应的效果（见图 9-8）。从图 9-7 与图 9-8 对比中发现，二者头部基本上一样的，都有一个返回按钮和一个提示性信息，所以很自然就会想到要进行布局的重用。

图 9-7 通知公告模块效果

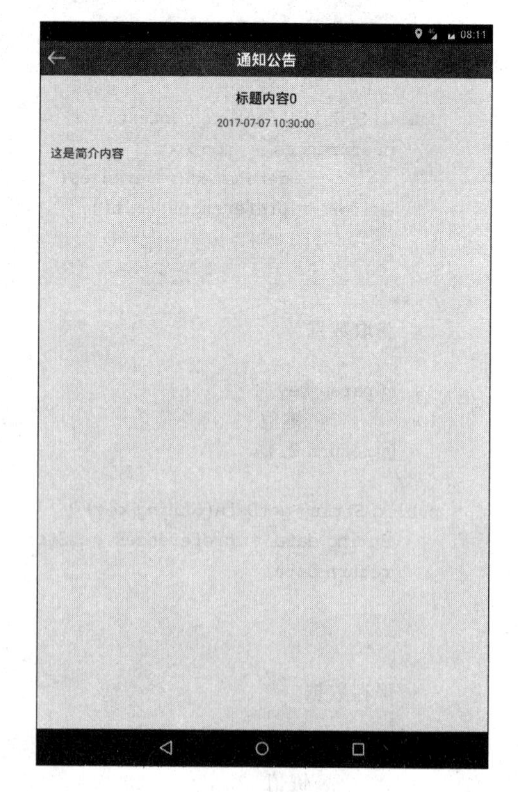

图 9-8 效果图

9.3.1 通知公告列表

在 res/layout 下新建一个 view_title. xml，是布局重用文件，里面用到的资源文件如表 9-4 所示。

表 9-4 图片资源

图片名称	图 片	说 明
ic_back	←	

view_title. xml 布局比较简单，只有一个 ImageView 和一个 TextView，主要代码如下：

```
<?xml version = "1.0" encoding = "utf - 8"?>
< RelativeLayout xmlns:android = "http://schemas.android.com/apk/res/android"
    android:layout_width = "match_parent"
    android:layout_height = "45dp"
    android:background = "@color/bg_title" >

    < ImageView
        android:id = "@ + id/iv_black"
        android:layout_width = "wrap_content"
        android:layout_height = "match_parent"
        android:layout_alignParentLeft = "true"
        android:layout_centerVertical = "true"
        android:paddingLeft = "10dp"
        android:src = "@drawable/ic_back" />

    < TextView
        android:id = "@ + id/tv_title"
        android:layout_width = "wrap_content"
        android:layout_height = "wrap_content"
        android:layout_centerInParent = "true"
        android:textColor = "#ffffff"
        android:textSize = "20sp"
        android:textStyle = "bold" />

</RelativeLayout>
```

在 res/layout 下新建一个 activity_message. xml 文件，里面用< include/>标签作为布局引用。主要代码如下：

```
<?xml version = "1.0" encoding = "utf - 8"?>
< LinearLayout xmlns:android = "http://schemas.android.com/apk/res/android"
    android:layout_width = "match_parent"
    android:layout_height = "match_parent"
    android:background = "@color/bg_all"
    android:orientation = "vertical" >

    < include layout = "@layout/view_title" />
```

```
    <ListView
        android:id = "@ + id/lv"
        android:layout_width = "match_parent"
        android:layout_height = "match_parent"
        android:scrollbars = "none" >
    </ListView>

</LinearLayout>
```

代码简洁明了，除了复用了一个标题之外，只有一个 ListView 控件了。在 ListView 中添加 Item 布局，新建一个 item_message.xml，主要功能是用三个 TextView 显示标题内容、内容简介和发布时间。主要代码如下：

```
<?xml version = "1.0" encoding = "utf - 8"?>
<RelativeLayout xmlns:android = "http://schemas.android.com/apk/res/android"
    android:layout_width = "match_parent"
    android:layout_height = "match_parent"
    android:orientation = "vertical" >

    <LinearLayout
        android:layout_width = "match_parent"
        android:layout_height = "wrap_content"
        android:layout_toLeftOf = "@ + id/tv_time"
        android:orientation = "vertical"
        android:padding = "10dp" >

        <TextView
            android:id = "@ + id/tv_titletext"
            android:layout_width = "match_parent"
            android:layout_height = "wrap_content"
            android:ellipsize = "end"
            android:singleLine = "true"
            android:textSize = "18sp" />

        <TextView
            android:id = "@ + id/tv_content"
            android:layout_width = "match_parent"
            android:layout_height = "wrap_content"
            android:layout_marginTop = "10dp"
            android:ellipsize = "end"
            android:maxLines = "2"
            android:textColor = "#99000000"
            android:textSize = "16sp" />
    </LinearLayout>

    <TextView
        android:id = "@ + id/tv_time"
        android:layout_width = "wrap_content"
        android:layout_height = "wrap_content"
        android:layout_alignParentBottom = "true"
```

```
        android:layout_alignParentRight = "true"
        android:text = "2016 - 04 - 23\n19:00:00"
        android:textColor = "@color/bg_title"
        android:gravity = "center"
        android:layout_centerVertical = "true"
        android:paddingRight = "10dp" />

</RelativeLayout>
```

在 com. qsd. kqxt. ctl 目录下新建 MessageAdapter. java 文件，继承 BaseAdapter 并实现其中方法，主要代码如下：

```
package com.qsd.kqxt.ctl;

import java.util.List;

import android.content.Context;
import android.view.View;
import android.view.ViewGroup;
import android.widget.BaseAdapter;
import android.widget.TextView;

import com.qsd.kqxt.R;
import com.qsd.kqxt.bean.MessageInfo;

public class MessageAdapter extends BaseAdapter {

    private List<MessageInfo> mInfos;
    private Context mContext;

    public MessageAdapter(Context context, List<MessageInfo> infos) {
        this.mContext = context;
        this.mInfos = infos;
    }

    @Override
    public int getCount() {
        return mInfos.size();
    }

    @Override
    public Object getItem(int arg0) {
        return null;
    }

    @Override
    public long getItemId(int position) {
        return mInfos.get(position).getId();
    }

    public class ViewHolder {
```

```
        // UI
        TextView mtitleTextView, mcontentTextView, mtimeTextView;
    }

    @Override
    public View getView(int position, View convertView, ViewGroup parent) {
        View view = convertView;
        ViewHolder holder = null;
        if (convertView == null) {
            view = View.inflate(mContext, R.layout.item_message, null);
            holder = new ViewHolder();

            holder.mtitleTextView = (TextView) view
                    .findViewById(R.id.tv_titletext);
            holder.mcontentTextView = (TextView) view
                    .findViewById(R.id.tv_content);
            holder.mtimeTextView = (TextView) view.findViewById(R.id.tv_time);

            view.setTag(holder);
        } else {
            holder = (ViewHolder) view.getTag();
        }

        holder.mtitleTextView.setText(mInfos.get(position).getNoticeTitle());
        holder.mcontentTextView.setText(mInfos.get(position)
                .getContentSummary());
        String dateAndTime = mInfos.get(position).getAddDate();
        String date = dateAndTime.substring(0, 10);
        String time = dateAndTime.substring(11, 18);
        dateAndTime = date + "\n" + time;
        holder.mtimeTextView.setText(dateAndTime);
        return view;
    }

}
```

新建一个 MessageInfo.java 实体类并进行封装，主要代码如下：

```
package com.qsd.kqxt.bean;

public class MessageInfo {
    private int id;
    private String noticeTitle;              // 标题
    private String contentSummary;           // 内容
    private String addDate;                  // 发布时间

    public int getId() {
        return id;
    }

    public void setId(int id) {
```

```
        this.id = id;
    }

    public String getNoticeTitle() {
        return noticeTitle;
    }

    public void setNoticeTitle(String noticeTitle) {
        this.noticeTitle = noticeTitle;
    }

    public String getContentSummary() {
        return contentSummary;
    }

    public void setContentSummary(String contentSummary) {
        this.contentSummary = contentSummary;
    }

    public String getAddDate() {
        return addDate;
    }

    public void setAddDate(String addDate) {
        this.addDate = addDate;
    }

}
```

在 com. qsd. kqxt. ui 目录下新建一个 MessageActivity. java，代码如下：

```
package com.qsd.kqxt.ui;

import java.util.ArrayList;
import java.util.List;

import android.app.Activity;
import android.content.Intent;
import android.os.Bundle;
import android.view.View;
import android.view.View.OnClickListener;
import android.widget.AdapterView;
import android.widget.AdapterView.OnItemClickListener;
import android.widget.ImageView;
import android.widget.ListView;
import android.widget.TextView;

import com.qsd.kqxt.R;
import com.qsd.kqxt.adapter.MessageAdapter;
import com.qsd.kqxt.bean.MessageInfo;
```

```
public class MessageActivity extends Activity implements OnClickListener {

    private TextView mtitleTextView;
    private ImageView mBlack;

    private ListView lstv;

    private MessageAdapter adapter;
    private List<MessageInfo> infos = new ArrayList<MessageInfo>();

    @Override
    protected void onCreate(Bundle savedInstanceState) {
        // TODO Auto-generated method stub
        super.onCreate(savedInstanceState);
        setContentView(R.layout.activity_message);

        initView();
        initData();
    }

    /***
     * 初始化布局控件
     */
    private void initView() {

        mtitleTextView = (TextView) findViewById(R.id.tv_title);
        mBlack = (ImageView) findViewById(R.id.iv_black);
        mBlack.setOnClickListener(this);
        mtitleTextView.setText("通知公告");
        lstv = (ListView) findViewById(R.id.lv);
        adapter = new MessageAdapter(this, infos);
        lstv.setAdapter(adapter);
        lstv.setOnItemClickListener(new OnItemClickListener() {

            @Override
            public void onItemClick(AdapterView<?> arg0, View arg1, int arg2,
             long arg3) {
                Intent intent = new Intent(MessageActivity.this,
                 MessageDetailActivity.class);
                intent.putExtra("id", infos.get(arg2).getId());
                intent.putExtra("content", infos.get(arg2)
                    .getContentSummary());
                intent.putExtra("title", infos.get(arg2).getNoticeTitle());
                intent.putExtra("time", infos.get(arg2).getAddDate());
                startActivity(intent);
            }
        });
    }

    /***
     * 初始化数据
```

```
     */
    private void initData() {
        for (int i = 0; i < 10; i++) {
            MessageInfo info = new MessageInfo();
            info.setId(i);
            info.setNoticeTitle("标题内容" + i);
            info.setContentSummary("这是简介内容");
            info.setAddDate("2017-07-07 10:30:00");
            infos.add(info);
        }

        adapter.notifyDataSetChanged();
    }

    @Override
    public void onClick(View v) {
        // TODO Auto-generated method stub
        switch (v.getId()) {
        case R.id.iv_black:// 返回
            this.finish();
            break;
        }
    }

}
```

9.3.2 通知公告详情

通知公告详情页面主要是显示通知或公告的全部信息，并对传过来的数据进行相应处理。布局比较简单：一行显示标题，一行显示时间，最后多行显示具体发布内容。因为是多行，内容有可能会超出屏幕，所以要在整个布局上加一个 ScrollView 控件。先新建一个 activity_message_detail.xml，主要代码如下：

```
<?xml version="1.0" encoding="utf-8"?>
<LinearLayout xmlns:android="http://schemas.android.com/apk/res/android"
    android:layout_width="match_parent"
    android:layout_height="match_parent"
    android:background="@color/bg_all"
    android:orientation="vertical" >

    <include layout="@layout/view_title" />

    <ScrollView
        android:layout_width="match_parent"
        android:layout_height="match_parent" >

        <LinearLayout
            android:layout_width="match_parent"
            android:layout_height="wrap_content"
            android:orientation="vertical"
```

```
                android:padding = "5dp" >

                <TextView
                    android:id = "@ + id/tv_titletext"
                    android:layout_width = "match_parent"
                    android:layout_height = "wrap_content"
                    android:layout_marginTop = "10dp"
                    android:ellipsize = "end"
                    android:gravity = "center_horizontal"
                    android:singleLine = "true"
                    android:textSize = "18sp"
                    android:text = "暂无数据" />

                <TextView
                    android:id = "@ + id/tv_time"
                    android:layout_width = "wrap_content"
                    android:layout_height = "wrap_content"
                    android:layout_gravity = "center_horizontal"
                    android:layout_marginTop = "10dp"
                    android:textColor = "@color/bg_title" />

                <TextView
                    android:id = "@ + id/tv_content"
                    android:layout_width = "match_parent"
                    android:layout_height = "wrap_content"
                    android:layout_marginTop = "10dp"
                    android:ellipsize = "end"
                    android:textColor = " #99000000"
                    android:textSize = "16sp"
                    android:gravity = "left|top"
                    android:padding = "10dp" />
            </LinearLayout>
        </ScrollView>

</LinearLayout>
```

在 com. qsd. kqxt. ui 目录下新建一个 MessageDetailActivity. java，主要功能是显示获取到的数据。代码如下：

```
package com.qsd.kqxt.ui;

import android.app.Activity;
import android.os.Bundle;
import android.view.View;
import android.view.View.OnClickListener;
import android.widget.ImageView;
import android.widget.TextView;

import com.qsd.kqxt.R;
import com.qsd.kqxt.bean.MessageContentInfo;
```

```
public class MessageDetailActivity extends Activity implements
OnClickListener {

    private TextView mtitleTextView;
    private TextView mMsgTitls;
    private TextView mTime;
    private TextView mContent;
    private ImageView mBlack;

    MessageContentInfo info;

    @Override
    protected void onCreate(Bundle savedInstanceState) {
        super.onCreate(savedInstanceState);
        setContentView(R.layout.activity_message_detail);
        initView();
        initData();
        getData();

    }

    /***
     * 初始化控件
     */
    private void initView() {

        mtitleTextView = (TextView) findViewById(R.id.tv_title);

        mMsgTitls = (TextView) findViewById(R.id.tv_titletext);

        mTime = (TextView) findViewById(R.id.tv_time);

        mContent = (TextView) findViewById(R.id.tv_content);
        mBlack = (ImageView) findViewById(R.id.iv_black);
        mBlack.setOnClickListener(this);
    }

    /***
     * 初始化数据
     */
    private void initData() {
        mtitleTextView.setText("通知公告");
        info = new MessageContentInfo();
        info.setId(getIntent().getIntExtra("id", 0));
        info.setAddDate(getIntent().getStringExtra("time"));
        info.setNoHtmlContent(getIntent().getStringExtra("content"));
        info.setNoticeTitle(getIntent().getStringExtra("title"));
    }

    /***
     * 获取数据并赋值
```

```
    */
    private void getData() {
        if (info != null) {
            mMsgTitls.setText(info.getNoticeTitle());
            mTime.setText(info.getAddDate());
            mContent.setText(info.getNoHtmlContent());
        }
    }

    @Override
    public void onClick(View v) {
        // TODO Auto-generated method stub
        switch (v.getId()) {
        case R.id.iv_black:// 返回
            this.finish();
            break;
        }
    }
}
```

9.4 工作日志模块

此模块主要功能是填写并提交工作日志，包含的信息有工作日志标题、工作日志内容填写、拍照功能、地点获取等。设计效果如图 9-9 所示，布局较为简单，包括一个头部、一个填写工作日志的区域、一个拍照按钮和一个地点信息显示区域。

图 9-9 工作日志模块效果

把图片放到 res/drawable-hdpi 目录下，其中用到的资源图片如表 9-5 所示。

表 9-5 图片资源

图片名称	图 片	说 明
ic_img_add	+	拍照的按钮图标

9.4.1 工作内容

资源准备完毕，在 res/layout 目录下编写 activity_work.xml 代码，实现工作日志页面的头部、工作日志填写区域。其中，GridView 实现上传多张拍摄的照片以及地点获取的页面显示。主要代码如下：

```
<?xml version = "1.0" encoding = "utf - 8"?>
< LinearLayout xmlns:android = "http://schemas.android.com/apk/res/android"
    android:id = "@ + id/main"
    android:layout_width = "match_parent"
    android:layout_height = "match_parent"
    android:background = "@color/bg_all"
    android:orientation = "vertical" >

    < RelativeLayout
        android:layout_width = "match_parent"
        android:layout_height = "wrap_content" >

        < include layout = "@layout/view_title" />

        < TextView
            android:id = "@ + id/tv_submit"
            android:layout_width = "wrap_content"
            android:layout_height = "45dp"
            android:layout_alignParentRight = "true"
            android:layout_centerVertical = "true"
            android:enabled = "false"
            android:padding = "10dp"
            android:paddingRight = "10dp"
            android:text = "提交"
            android:textColor = "#ffffff"
            android:textSize = "18sp" />
    </RelativeLayout >

    < RelativeLayout
        android:layout_width = "match_parent"
        android:layout_height = "200dp"
        android:background = "#ffffff"
        android:padding = "10dp" >

        < EditText
            android:id = "@ + id/edt_work"
```

```
            android:layout_width = "match_parent"
            android:layout_height = "wrap_content"
            android:background = "@null"
            android:hint = "请填写工作日志,不得少于 50 个汉字" />
    </RelativeLayout>

    <GridView
        android
        android:scrollbars = "none"
        android:verticalSpacing = "20dp" >
    </GridView>

    <LinearLayout
        android:layout_width = "wrap_content"
        android:layout_height = "wrap_content":id = "@ + id/gv_images"
        android:layout_width = "wrap_content"
        android:layout_height = "wrap_content"
        android:background = " # fff"
        android:columnWidth = "100dp"
        android:horizontalSpacing = "20dp"
        android:numColumns = "3"

        android:background = " # ffffff" >

        <View
            android:layout_width = "match_parent"
            android:layout_height = "1dp"
            android:layout_alignParentBottom = "true"
            android:layout_marginLeft = "10dp"
            android:layout_marginRight = "10dp"
            android:background = " # 66000000" />
    </LinearLayout>

    <RelativeLayout
        android:layout_width = "match_parent"
        android:layout_height = "0dp"
        android:layout_weight = "1" >

        <TextView
            android:id = "@ + id/tv_location"
            android:layout_width = "match_parent"
            android:layout_height = "wrap_content"
            android:background = " # ffffff"
            android:padding = "10dp"
            android:text = "地点: 正在获取中..." >
        </TextView>
    </RelativeLayout>

</LinearLayout>
```

在 com. qsd. kqxt. bean 包下新建 LocationInfo .java,该类在 Location 工具类中将被调

用。主要代码如下：

```
package com.qsd.kqxt.bean;

public class LocationInfo {

    String latitude;                          // 纬度
    String longitude;                         // 经度
    String add;
    String locDescribution;
    String city;

    public String getCity() {
        return city;
    }

    public void setCity(String city) {
        this.city = city;
    }

    public String getLatitude() {
        return latitude;
    }

    public void setLatitude(String latitude) {
        this.latitude = latitude;
    }

    public String getLongitude() {
        return longitude;
    }

    public void setLongitude(String longitude) {
        this.longitude = longitude;
    }

    public String getAdd() {
        return add;
    }

    public void setAdd(String add) {
        this.add = add;
    }

    public String getLocDescribution() {
        return locDescribution;
    }

    public void setLocDescribution(String locDescribution) {
        this.locDescribution = locDescribution;
    }

}
```

在 com. qsd. kqxt. utils 包下新建 Location. Java，该类可实现地点定位。主要代码如下：

```
package com.qsd.kqxt.utils;

import android.content.Context;
import android.os.Handler;
import android.os.Message;
import android.util.Log;

import com.baidu.location.BDLocation;
import com.baidu.location.BDLocationListener;
import com.baidu.location.LocationClient;
import com.baidu.location.LocationClientOption;
import com.baidu.location.LocationClientOption.LocationMode;
import com.qsd.kqxt.bean.LocationInfo;

public class Location {

    Handler handler;
    public LocationClient mLocationClient = null;
    public BDLocationListener myListener;
    LocationInfo info;
    Context context;

    public Location(Handler handler, Context context) {
        this.handler = handler;
        this.context = context;
        info = new LocationInfo();
        myListener = new MyLocationListener();
        //初始化 LocationClient 类
        mLocationClient = new LocationClient(context);          // 声明 LocationClient 类
        mLocationClient.registerLocationListener(myListener);   // 注册监听函数
        initLocation();
    }

    //开始定位
    public void start() {
        mLocationClient.start();
    }

    //停止定位
    public void stop() {
        mLocationClient.stop();
    }

    //配置定位 SDK 参数
    private void initLocation() {
        LocationClientOption option = new LocationClientOption();
        option.setLocationMode(LocationMode.Hight_Accuracy);
```

```
    //设置定位模式,可选,默认为高精度,这里设置为高精度、低功耗、仅设备

    option.setCoorType("bd09ll");
    //设置返回的定位结果坐标系,可选,默认为 gcj02,这里设为 bd09ll
    // gcj02 为火星坐标系,bd09ll 为百度坐标系

    int span = 0;
    option.setScanSpan(span);
    //设置发起定位请求的间隔,可选,默认为 0,即仅定位一次
    //如果需要连续定位,则需要大于等于 1000ms 方有效

    option.setIsNeedAddress(true);
    //设置是否需要地址信息,可选,默认不需要(false)

    option.setOpenGps(true);
    //设置是否使用 GPS,可选,默认不使用(false)

    option.setLocationNotify(true);
    //设置是否当 GPS 有效时按照 1 次每秒的频率输出 GPS 结果,可选,默认不输出(false)

    option.setIsNeedLocationDescribe(true);
    //设置是否需要位置语义化结果(结果类似于"在长春国贸大厦附近")
    //可选,默认不需要(false),可以在 BDLocation.getLocationDescribe 里得到

    option.setIsNeedLocationPoiList(true);
    // 设置是否需要 POI 结果,可选,默认不需要(false)
    // 可以在 BDLocation.getPoiList 里得到
    option.setIgnoreKillProcess(false);
    // 设置是否在 stop 的时候杀死这个进程,可选,默认不杀死(true)
    // 定位在 SDK 内部是一个 SERVICE,并被放到了独立进程中

    option.SetIgnoreCacheException(false);
    //设置是否收集 CRASH 信息,可选,默认收集(false)

    option.setEnableSimulateGps(false);
    //设置是否需要过滤 GPS 仿真结果,可选,默认需要(false)

    mLocationClient.setLocOption(option);
}

//实现 BDLocationListener 接口
public class MyLocationListener implements BDLocationListener {

    @Override
    public void onReceiveLocation(BDLocation location) {
        String latitude = String.valueOf(location.getLatitude());
        String longitude = String.valueOf(location.getLongitude());
        String add = location.getAddrStr();
        String locDescribution = location.getLocationDescribe();
        info.setLatitude(String.valueOf(location.getLatitude()));
        info.setLongitude(String.valueOf(location.getLongitude()));
```

```
                info.setAdd(location.getAddrStr());
                info.setLocDescribution(location.getLocationDescribe());
                info.setCity(location.getCity());
                Log.i("location", "经度: " + longitude + "\n 纬度: " + latitude + "\n
                        位置: "+ location + locDescribution);
                Message msg = new Message();
                msg.what = 0;
                msg.obj = info;
                handler.sendMessage(msg);
            }

            @Override
            public void onConnectHotSpotMessage(String arg0, int arg1) {
                // TODO Auto - generated method stub

            }

        }

    }
```

在 com.qsd.kqxt.ctl 包下新建 GridAdapter.java。该类对拍照上传图片进行图片大小的调整以适应手机屏幕的功能。主要代码如下：

```
package com.qsd.kqxt.ctl;

import java.util.List;

import android.content.Context;
import android.graphics.Bitmap;
import android.graphics.BitmapFactory;
import android.util.Log;
import android.view.View;
import android.view.ViewGroup;
import android.widget.BaseAdapter;
import android.widget.ImageView;

import com.qsd.kqxt.R;
import com.qsd.kqxt.utils.BitmapUtils;

public class GridAdapter extends BaseAdapter {

    private List < String > urls;
    Context context;

    public GridAdapter(List < String > urls, Context context) {
        this.urls = urls;
        this.context = context;
    }

    @Override
```

```
public int getCount() {
    return urls.size();
}

@Override
public Object getItem(int position) {
    return null;
}

@Override
public long getItemId(int position) {
    return 0;
}

@Override
public View getView(int position, View convertView, ViewGroup parent) {
    ViewHolder viewHolder;
    View view;
    if (convertView == null) {
        view = View.inflate(context, R.layout.item_gridview, null);
        viewHolder = new ViewHolder();
        viewHolder.mImageView = (ImageView) view
         .findViewById(R.id.iv_image_item);
        view.setTag(viewHolder);
    } else {
        view = convertView;
        viewHolder = (ViewHolder) convertView.getTag();
    }
    Log.e("adapter", position + "");

    if (position == 0) {
        viewHolder.mImageView.setImageResource(R.drawable.ic_img_add);
    } else {
        Log.e("adapter__1", urls.size() + "");
        Log.i("Uri---", urls.get(position));
        //图片太大会不清晰,所以要提高分辨率或者修改图片大小
        BitmapFactory.Options options = new BitmapFactory.Options();
        options.inJustDecodeBounds = true;
        BitmapFactory.decodeFile(urls.get(position), options);
        options.inSampleSize = 3;
        options.inJustDecodeBounds = false;
        Bitmap bitmap = BitmapFactory.decodeFile(urls.get(position),
         options);
        viewHolder.mImageView.setImageBitmap(bitmap);
    }
    return view;
}

private class ViewHolder {
```

```
        ImageView mImageView;
    }

}
```

在 src/com.qsd.kqxt.ui 下编写 WorkActivity.java 代码，在代码中调用此 Java 方法，以实现文本区域文字的编辑、图片选择(相机拍照上传)、定位、设置标题内容、提交等功能，代码如下：

```
package com.qsd.kqxt.ui;

import java.io.File;
import java.util.ArrayList;
import java.util.List;

import android.app.Activity;
import android.content.Intent;
import android.net.Uri;
import android.os.Bundle;
import android.os.Environment;
import android.os.Handler;
import android.provider.MediaStore;
import android.view.View;
import android.view.View.OnClickListener;
import android.widget.AdapterView;
import android.widget.AdapterView.OnItemClickListener;
import android.widget.EditText;
import android.widget.GridView;
import android.widget.ImageView;
import android.widget.LinearLayout;
import android.widget.TextView;

import com.qsd.kqxt.R;
import com.qsd.kqxt.adapter.GridAdapter;
import com.qsd.kqxt.bean.LocationInfo;
import com.qsd.kqxt.utils.Location;

public class WorkActivity extends Activity {

    private TextView mtitleTextView;
    private EditText mEditText;
    private TextView mSubmit;
    private GridView mGridView;
    private LinearLayout mMain;
    private TextView mLocation;
    private ImageView mBlack;
    GridAdapter adapter;
    List<String> urls = new ArrayList<String>();
    Location location;
    String address;
    String loc;
    String lon;
    String lat;
```

```
File photoFile;

Handler handler = new Handler() {
    public void handleMessage(android.os.Message msg) {
        LocationInfo info = (LocationInfo) msg.obj;
        address = info.getAdd();
        loc = info.getLocDescribution();
        lon = info.getLongitude();
        lat = info.getLatitude();
        if (loc == null || loc.equals("") || lon == null || lon.equals("")
         || lat == null || lat.equals("")) {
            return;
        }
        if (address == null || address.equals("")) {
            return;
        }
        mLocation.setText("地点：" + address + loc);
    };
};

@Override
protected void onCreate(Bundle savedInstanceState) {
    super.onCreate(savedInstanceState);
    setContentView(R.layout.activity_work);
    initView();
    initData();
    setView();

}

private void initView() {
    // TODO Auto-generated method stub
    mtitleTextView = (TextView) findViewById(R.id.tv_title);
    mEditText = (EditText) findViewById(R.id.edt_work);
    mSubmit = (TextView) findViewById(R.id.tv_submit);

    mGridView = (GridView) findViewById(R.id.gv_images);
    mMain = (LinearLayout) findViewById(R.id.main);
    mLocation = (TextView) findViewById(R.id.tv_location);
    mBlack = (ImageView) findViewById(R.id.iv_black);

    mBlack.setOnClickListener(new OnClickListener() {
        @Override
        public void onClick(View v) {
            finish();
        }
    });
    mSubmit.setOnClickListener(new OnClickListener() {
        @Override
        public void onClick(View v) {
            finish();
        }
    });
}
```

```
    private void initData() {
        location = new Location(handler, WorkActivity.this);
        location.start();
        mtitleTextView.setText("工作日志");
        urls.add("R.drawable.ic_img_add");
        adapter = new GridAdapter(urls, WorkActivity.this);
    }

    private void setView() {
        mGridView.setAdapter(adapter);
        mGridView.setOnItemClickListener(new OnItemClickListener() {
            @Override
            public void onItemClick(AdapterView<?> parent, View view,
                    int position, long id) {
                if (position == 0) {
                    mEditText.clearFocus();
                    toCamere();
                } else {
                    urls.remove(position);
                    adapter.notifyDataSetChanged();
                }
            }
        });
    }


/***
 * 调取照相机功能
 */
    private void toCamere() {
        File file = new File(Environment.getExternalStorageDirectory() +
"/Pic");
        if (!file.exists()) {
            file.mkdir();
        }
        // 图片路径。如果不这样设置,获取的照片会是系统照相机的缩略图
        photoFile = new File(Environment.getExternalStorageDirectory()
                + "/Pic/" + System.currentTimeMillis() + ".jpg");
        Intent intent = new Intent(MediaStore.ACTION_IMAGE_CAPTURE);
        intent.putExtra(MediaStore.EXTRA_OUTPUT, Uri.fromFile(photoFile));
        startActivityForResult(intent, 0);
    }

    protected void onActivityResult(int requestCode, int resultCode, Intent data) {
        super.onActivityResult(requestCode, resultCode, data);
        if (resultCode == RESULT_OK) {
            if (requestCode == 0) {
                urls.add(photoFile.getAbsolutePath());
                adapter.notifyDataSetChanged();
            }

        }
    }
```

```
    /***
     * 提交的一些操作,例如验证是否为空等操作
     */
    public void submit() {
        this.finish();
    }

}
```

9.4.2 图片选择

在 src/com.qsd.kqxt.activity 下的 WorkActivity.java 中,选择图片的主要代码如下：

```
private void setView() {
        mGridView.setAdapter(adapter);
        mGridView.setOnItemClickListener(new OnItemClickListener() {
            @Override
            public void onItemClick(AdapterView<?> parent, View view,
                    int position, long id) {
                if (position == 0) {
                    mEditText.clearFocus();
                    toCamere();
                } else {
                    urls.remove(position);
                    adapter.notifyDataSetChanged();
                }
            }
        });
    }

/***
 * 调取照相机功能
 */
    private void toCamere() {
        File file = new File(Environment.getExternalStorageDirectory() + "/Pic");
        if (!file.exists()) {
            file.mkdir();
        }
        // 图片路径。如果不这样设置,获取的照片会是系统照相机的缩略图
        photoFile = new File(Environment.getExternalStorageDirectory()
                + "/Pic/" + System.currentTimeMillis() + ".jpg");
        Intent intent = new Intent(MediaStore.ACTION_IMAGE_CAPTURE);
        intent.putExtra(MediaStore.EXTRA_OUTPUT, Uri.fromFile(photoFile));
        startActivityForResult(intent, 0);
    }
```

9.4.3 定位

在 src/com.qsd.kqxt.activity 下的 WorkActivity.java 中,定位的主要代码如下：

```
Handler handler = new Handler() {
```

```
        public void handleMessage(android.os.Message msg) {
            LocationInfo info = (LocationInfo) msg.obj;
            address = info.getAdd();
            loc = info.getLocDescribution();
            lon = info.getLongitude();
            lat = info.getLatitude();
            if (loc == null || loc.equals("") || lon == null || lon.equals("")
                || lat == null || lat.equals("")) {
                return;
            }
            if (address == null || address.equals("")) {
                return;
            }
            mLocation.setText("地点: " + address + loc);
        };
    };
```

9.5 考勤管理模块

此模块主要功能是获取定位信息以及签到。设计效果如图 9-10 所示，布局为一个头部和定位信息以及签到按钮。

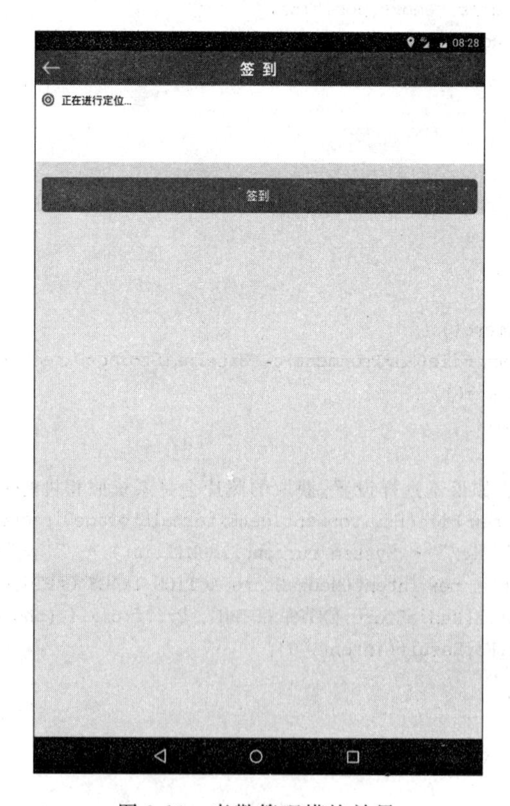

图 9-10　考勤管理模块效果

把图片放到 res/drawable-hdpi 目录下，其中用到的资源图片如表 9-6 所示。

表 9-6　图片资源

图片名称	图　片	说　明
radiobg_press	◎	

定位与签到(每天只能提交一次)

在 res/layout 目录下添加 activity_check. xml 文件，通过 include 引入标题部分，并实现定位区域以及签到的区域的页面显示。主要代码如下：

```
<?xml version = "1.0" encoding = "utf - 8"?>
<LinearLayout xmlns:android = "http://schemas.android.com/apk/res/android"
    android:layout_width = "match_parent"
    android:layout_height = "match_parent"
    android:background = "@color/bg_all"
    android:orientation = "vertical" >

    <include layout = "@layout/view_title" />

    <LinearLayout
        android:layout_width = "match_parent"
        android:layout_height = "wrap_content"
        android:background = "#ffffff"
        android:orientation = "vertical" >

        <TextView
            android:id = "@ + id/tv_location"
            android:layout_width = "match_parent"
            android:layout_height = "wrap_content"
            android:background = "#ffffff"
            android:drawableLeft = "@drawable/ic_local"
            android:drawablePadding = "10dp"
            android:gravity = "center_vertical"
            android:padding = "10dp"
            android:text = "正在进行定位..." >
        </TextView>

        <LinearLayout
            android:id = "@ + id/ll_popup"
            android:layout_width = "match_parent"
            android:layout_height = "wrap_content"
            android:background = "#ffeeeeee"
            android:orientation = "vertical" >

            <RelativeLayout
                android:layout_width = "match_parent"
                android:layout_height = "60dp"
```

```
                android:background = "#fff" >

                <TextView
                    android:id = "@ + id/popupwindow_calendar_month"
                    android:layout_width = "match_parent"
                    android:layout_height = "60dp"
                    android:gravity = "center"
                    android:textColor = "#000"
                    android:textSize = "18sp" />
            </RelativeLayout>


            <Button
                android:id = "@ + id/btn_signIn"
                android:layout_width = "match_parent"
                android:layout_height = "45dp"
                android:layout_marginLeft = "10dp"
                android:layout_marginRight = "10dp"
                android:layout_marginTop = "20dp"
                android:background = "@drawable/btn_sign_bg_selector"
                android:text = "签到"
                android:textColor = "@android:color/white"
                android:textSize = "16sp" />
        </LinearLayout>
    </LinearLayout>

</LinearLayout>
```

在 com. qsd. kqxt. ui 包下新建 CheckActivity 类,实现定位和签到功能。主要代码如下:

```
package com.qsd.kqxt.ui;

import com.qsd.kqxt.R;
import com.qsd.kqxt.bean.LocationInfo;
import com.qsd.kqxt.utils.Location;

import android.app.Activity;
import android.os.Bundle;
import android.os.Handler;
import android.view.View;
import android.view.View.OnClickListener;
import android.widget.Button;
import android.widget.ImageView;
import android.widget.TextView;

public class CheckActivity extends Activity implements OnClickListener {

    private TextView mtitleTextView;
    private Button mSign;
    private TextView mLocation;
```

```
private ImageView mBlack;
Location location;
String address;
String loc;
String lon;
String lat;
String latitude;            // 纬度
String longitude;           // 经度
String add;
String locDescribution;

Handler handler = new Handler() {
    public void handleMessage(android.os.Message msg) {
        LocationInfo info = (LocationInfo) msg.obj;
        latitude = info.getLatitude();
        longitude = info.getLongitude();
        add = info.getAdd();
        locDescribution = info.getLocDescribution();
        if (locDescribution == null || locDescribution.equals("")
                || longitude == null || longitude.equals("")
                || latitude == null || latitude.equals("")) {
            return;
        }
        if (add == null || add.equals("")) {
            return;
        }
        mLocation.setText(add + locDescribution);
    };
};

@Override
protected void onCreate(Bundle savedInstanceState) {
    super.onCreate(savedInstanceState);
    setContentView(R.layout.activity_check);
    initView();
    initData();
}

private void initView() {
    mtitleTextView = (TextView) findViewById(R.id.tv_title);
    mSign = (Button) findViewById(R.id.btn_signIn);
    mLocation = (TextView) findViewById(R.id.tv_location);
    mBlack = (ImageView) findViewById(R.id.iv_black);
    mBlack.setOnClickListener(this);
    mSign.setOnClickListener(this);
}

private void initData() {
    mtitleTextView.setText("签 到");
    location = new Location(handler, CheckActivity.this);
    location.start();
```

```
        }

        @Override
        public void onClick(View v) {
            // TODO Auto-generated method stub
            switch (v.getId()) {
            case R.id.iv_black:// 返回
                this.finish();
                break;
            case R.id.btn_signIn:// 签到
                this.finish();
                break;
            }
        }

    }
```

9.6 费用申请模块

此模块包括两个页面：费用申请页面以及费用审批页面。费用申请页面主要填写费用申请信息以及提交费用申请，如图 9-11 所示。费用审批页面主要展示费用申请的相关信息，包括费用审批的标题内容，各个费用申请条目的费用名称、申请人、申请时间以及审批状态，如图 9-12 所示。

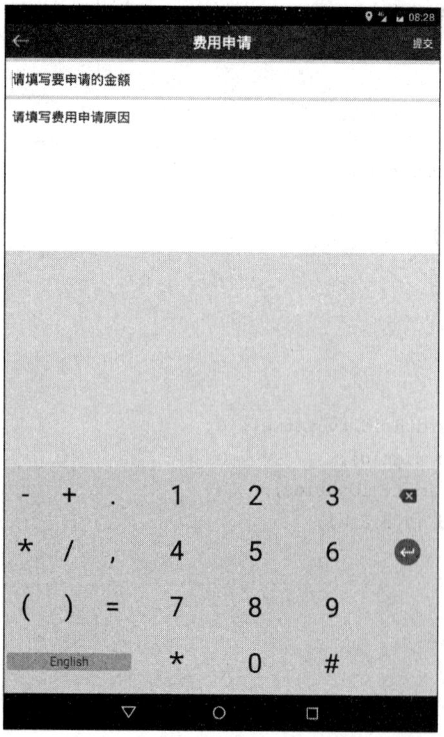

图 9-11 费用申请页面效果

图 9-12 费用审批页面效果

9.6.1 费用审批列表

在 res/layout 目录下新建 item_cost. xml 文件，实现费用审批页面每一个费用条目的页面显示，包括费用名称、申请人、申请时间和审批状态。主要代码如下：

```
<?xml version = "1.0" encoding = "utf - 8"?>
< LinearLayout xmlns:android = "http://schemas.android.com/apk/res/android"
    android:layout_width = "match_parent"
    android:layout_height = "wrap_content"
    android:orientation = "horizontal" >

    < LinearLayout
        android:layout_width = "0dp"
        android:layout_height = "wrap_content"
        android:layout_weight = "5"
        android:orientation = "vertical"
        android:padding = "10dp" >

        < LinearLayout
            android:layout_width = "match_parent"
            android:layout_height = "wrap_content"
            android:orientation = "horizontal" >

            < TextView
                android:layout_width = "wrap_content"
                android:layout_height = "wrap_content"
                android:ellipsize = "end"
                android:singleLine = "true"
                android:text = "费用名称: "
                android:textSize = "18sp" />

            < TextView
                android:id = "@ + id/tv_name"
                android:layout_width = "match_parent"
                android:layout_height = "wrap_content"
                android:ellipsize = "end"
                android:singleLine = "true"
                android:textColor = "@color/bg_title"
                android:textSize = "18sp" />
        </LinearLayout >

        < LinearLayout
            android:layout_width = "match_parent"
            android:layout_height = "wrap_content"
            android:orientation = "horizontal" >

            < TextView
                android:layout_width = "wrap_content"
                android:layout_height = "wrap_content"
```

```
                android:ellipsize = "end"
                android:singleLine = "true"
                android:text = "申请人 ："
                android:textSize = "18sp" />

            <TextView
                android:id = "@ + id/tv_person"
                android:layout_width = "match_parent"
                android:layout_height = "wrap_content"
                android:ellipsize = "end"
                android:singleLine = "true"
                android:textColor = "@color/bg_title"
                android:textSize = "18sp" />
        </LinearLayout>

        <LinearLayout
            android:layout_width = "match_parent"
            android:layout_height = "wrap_content"
            android:orientation = "horizontal" >

            <TextView
                android:layout_width = "wrap_content"
                android:layout_height = "wrap_content"
                android:ellipsize = "end"
                android:singleLine = "true"
                android:text = "申请时间："
                android:textSize = "18sp" />

            <TextView
                android:id = "@ + id/tv_time"
                android:layout_width = "match_parent"
                android:layout_height = "wrap_content"
                android:ellipsize = "end"
                android:singleLine = "true"
                android:textColor = "@color/bg_title"
                android:textSize = "18sp" />
        </LinearLayout>
    </LinearLayout>

    <TextView
        android:id = "@ + id/tv_state"
        android:layout_width = "0dp"
        android:layout_height = "wrap_content"
        android:layout_gravity = "center_vertical"
        android:layout_marginLeft = "5dp"
        android:layout_marginRight = "15dp"
        android:layout_weight = "2"
        android:background = "@color/bg_title"
        android:gravity = "center"
        android:paddingBottom = "5dp"
        android:paddingLeft = "10dp"
```

```
        android:paddingRight = "10dp"
        android:paddingTop = "5dp"
        android:text = "审批中"
        android:textColor = "#ffffff" />

</LinearLayout>
```

在 res/layout 目录下新建 activity_cost. xml 文件，实现费用审批页面的头部信息(include 标签)、费用申请按钮与 items 整体区域的显示。主要代码如下：

```
<?xml version = "1.0" encoding = "utf - 8"?>
<LinearLayout xmlns:android = "http://schemas.android.com/apk/res/android"
    android:layout_width = "match_parent"
    android:layout_height = "match_parent"
    android:background = "@color/bg_all"
    android:orientation = "vertical" >

    <RelativeLayout
        android:layout_width = "match_parent"
        android:layout_height = "wrap_content" >

        <include layout = "@layout/view_title" />

        <TextView
            android:id = "@ + id/tv_apply"
            android:layout_width = "wrap_content"
            android:layout_height = "wrap_content"
            android:layout_alignParentRight = "true"
            android:layout_centerVertical = "true"
            android:paddingRight = "10dp"
            android:text = "费用\n 申请"
            android:textColor = "#ffffff" />
    </RelativeLayout>

    <ListView
        android:id = "@ + id/lstv"
        android:layout_width = "match_parent"
        android:layout_height = "match_parent"
        android:background = "#fff"
        android:divider = "#F7F7F7"
        android:dividerHeight = "5dp"
        android:scrollbars = "none" >
    </ListView>

</LinearLayout>
```

在 com. qsd. kqxt. bean 包下新建 CostInfo 类。在 CostAdapter 中将调用此类。主要代码如下：

```
package com.qsd.kqxt.bean;
```

```
public class CostInfo {
    private int id;
    private String feeName;
    private String staffName;
    private String feeDate;
    private int approvalStatus;
    public int getId() {
        return id;
    }
    public void setId(int id) {
        this.id = id;
    }
    public String getFeeName() {
        return feeName;
    }
    public void setFeeName(String feeName) {
        this.feeName = feeName;
    }
    public String getStaffName() {
        return staffName;
    }
    public void setStaffName(String staffName) {
        this.staffName = staffName;
    }
    public String getFeeDate() {
        return feeDate;
    }
    public void setFeeDate(String feeDate) {
        this.feeDate = feeDate;
    }
    public int getApprovalStatus() {
        return approvalStatus;
    }
    public void setApprovalStatus(int approvalStatus) {
        this.approvalStatus = approvalStatus;
    }


}
```

在 com.qsd.kqxt.ctl 包下新建 CostAdapter 类,实现对 item_cost.xml 文件的多条显示。主要代码如下:

```
package com.qsd.kqxt.ctl;

import java.util.List;

import android.content.Context;
import android.view.View;
import android.view.ViewGroup;
import android.widget.BaseAdapter;
```

```
import android.widget.TextView;

import com.qsd.kqxt.R;
import com.qsd.kqxt.bean.CostInfo;

public class CostAdapter extends BaseAdapter {

    private List<CostInfo> mInfos;
    private Context mContext;

    public CostAdapter(Context context, List<CostInfo> infos) {
        this.mContext = context;
        this.mInfos = infos;
    }

    @Override
    public int getCount() {
        return mInfos.size();
    }

    @Override
    public Object getItem(int arg0) {
        return null;
    }

    @Override
    public long getItemId(int position) {
        Long id = Long.parseLong(mInfos.get(position).getId() + "");
        return id;
    }

    public class ViewHolder {
        // UI
        TextView mtitleTextView, mpersonTextView, mtimeTextView,
                mstateTextView;
    }

    @Override
    public View getView(int position, View convertView, ViewGroup parent) {
        View view = convertView;
        ViewHolder holder = null;
        if (convertView == null) {
            view = View.inflate(mContext, R.layout.item_cost, null);
            holder = new ViewHolder();

            holder.mtitleTextView = (TextView) view.findViewById(R.id.tv_name);
            holder.mpersonTextView = (TextView) view
                    .findViewById(R.id.tv_person);
            holder.mtimeTextView = (TextView) view.findViewById(R.id.tv_time);
            holder.mstateTextView = (TextView)
                view.findViewById(R.id.tv_state);
```

```
            view.setTag(holder);
        } else {
            holder = (ViewHolder) view.getTag();
        }

        holder.mtitleTextView.setText(mInfos.get(position).getFeeName());
        holder.mpersonTextView.setText(mInfos.get(position).getStaffName());
        holder.mtimeTextView.setText(mInfos.get(position).getFeeDate());
        if (mInfos.get(position).getApprovalStatus() == 0) {
            holder.mstateTextView.setText("审批中");
        } else if (mInfos.get(position).getApprovalStatus() == 1) {
            holder.mstateTextView.setText("审批完成");
        } else if (mInfos.get(position).getApprovalStatus() == 2) {
            holder.mstateTextView.setText("审批驳回");
        }
        return view;
    }

}
```

在 com.qsd.kqxt.activity 包下新建 CostActivity 类，实现调用 CostAdapter 实现多条信息展示、头部标题信息的设置、单击每个条目的序数显示，代码如下：

```
package com.qsd.kqxt.activity;

import java.util.ArrayList;
import java.util.List;

import com.qsd.kqxt.R;
import com.qsd.kqxt.adapter.CostAdapter;
import com.qsd.kqxt.bean.CostInfo;

import android.app.Activity;
import android.content.Intent;
import android.os.Bundle;
import android.view.View;
import android.view.View.OnClickListener;
import android.widget.AdapterView;
import android.widget.AdapterView.OnItemClickListener;
import android.widget.ImageView;
import android.widget.ListView;
import android.widget.TextView;
import android.widget.Toast;

public class CostActivity extends Activity implements OnClickListener {
    private TextView mtitleTextView;
    private TextView mApply;
    private ImageView mBlack;

    private ListView lstv;
```

```
private CostAdapter adapter;
private List<CostInfo> infos = new ArrayList<CostInfo>();

@Override
protected void onCreate(Bundle savedInstanceState) {
    // TODO Auto-generated method stub
    super.onCreate(savedInstanceState);
    setContentView(R.layout.activity_cost);

    initView();
    initData();
}

private void initData() {
    for (int i = 0; i < 10; i++) {
        CostInfo info = new CostInfo();
        info.setId(i);
        info.setApprovalStatus(i / 3);
        info.setFeeDate("2017-02-20");
        info.setStaffName("张三");
        info.setFeeName("办公用品");
        infos.add(info);
    }
    adapter = new CostAdapter(this, infos);
    lstv.setAdapter(adapter);
    lstv.setOnItemClickListener(new OnItemClickListener() {

        @Override
        public void onItemClick(AdapterView<?> arg0, View arg1, int arg2,
                long arg3) {
            // 可以将 Toast 单独写一个工具类
            Toast.makeText(getApplicationContext(), "单击了第" + arg2,
                    Toast.LENGTH_SHORT).show();
        }
    });
}

private void initView() {
    mtitleTextView = (TextView) findViewById(R.id.tv_title);
    mtitleTextView.setText("费用审批");
    mApply = (TextView) findViewById(R.id.tv_apply);
    mBlack = (ImageView) findViewById(R.id.iv_black);
    lstv = (ListView) findViewById(R.id.lstv);

    mApply.setOnClickListener(this);
    mBlack.setOnClickListener(this);

}

@Override
public void onClick(View v) {
```

```
            // TODO Auto - generated method stub
            switch (v.getId()) {
            case R.id.iv_black:// 返回
                this.finish();
                break;
            case R.id.tv_apply:
                Intent intent = new Intent(this, CostApplyActivity.class);
                startActivity(intent);
                break;
            }
        }

    }
```

9.6.2 费用申请

在 res/layout 目录下新建 activity_cost_apply.xml，实现费用申请页面的头部区域(include 标签引入)、填写申请金额、申请原因的区域显示。代码如下：

```
<?xml version = "1.0" encoding = "utf - 8"?>
<LinearLayout xmlns:android = "http://schemas.android.com/apk/res/android"
    android:layout_width = "match_parent"
    android:layout_height = "match_parent"
    android:background = "@color/bg_all"
    android:orientation = "vertical" >

    <RelativeLayout
        android:layout_width = "match_parent"
        android:layout_height = "wrap_content" >

        <include layout = "@layout/view_title" />

        <TextView
            android:id = "@ + id/tv_submit"
            android:layout_width = "wrap_content"
            android:layout_height = "wrap_content"
            android:layout_alignParentRight = "true"
            android:layout_centerVertical = "true"
            android:padding = "10dp"
            android:text = "提交"
            android:textColor = "#ffffff" />
    </RelativeLayout>

    <EditText
        android:id = "@ + id/edt_work"
        android:layout_width = "match_parent"
        android:layout_height = "45dp"
        android:layout_marginBottom = "5dp"
        android:layout_marginTop = "5dp"
        android:background = "#ffffff"
```

```
        android:hint = "请填写要申请的金额"
        android:numeric = "integer"
        android:padding = "10dp"
        android:textColor = "@color/btn_submit" />

    <RelativeLayout
        android:layout_width = "match_parent"
        android:layout_height = "200dp"
        android:background = "#ffffff"
        android:padding = "10dp" >

        <EditText
            android:id = "@ + id/edt_work_reason"
            android:layout_width = "match_parent"
            android:layout_height = "match_parent"
            android:background = "@null"
            android:gravity = "top|left"
            android:hint = "请填写费用申请原因" />
    </RelativeLayout>


</LinearLayout>
```

在 com. qsd. kqxt. ui 包下新建 CostApplyActivity 类。该类实现对用户提交的费用金额的数额的判断以及提交费用申请等功能。代码如下：

```
package com.qsd.kqxt.ui;

import com.qsd.kqxt.R;
import com.qsd.kqxt.bean.LocationInfo;

import android.app.Activity;
import android.os.Bundle;
import android.os.Handler;
import android.view.View;
import android.view.View.OnClickListener;
import android.widget.ImageView;
import android.widget.TextView;
import android.widget.Toast;

public class CostApplyActivity extends Activity implements OnClickListener {
    private TextView mtitleTextView;
    private TextView mMoney;
    private TextView mReason;
    private TextView mLocation;
    private TextView mSubmint;
    private ImageView mBlack;
    String add;
    String locDescribution;

    @Override
```

```
protected void onCreate(Bundle savedInstanceState) {
    // TODO Auto - generated method stub
    super.onCreate(savedInstanceState);
    setContentView(R.layout.activity_cost_apply);
    initView();
    initData();

}

private void initView() {
    // TODO Auto - generated method stub

    mtitleTextView = (TextView) findViewById(R.id.tv_title);
    mMoney = (TextView) findViewById(R.id.edt_work);
    mReason = (TextView) findViewById(R.id.edt_work_reason);
    mLocation = (TextView) findViewById(R.id.tv_location);
    mBlack = (ImageView) findViewById(R.id.iv_black);
    mSubmint = (TextView) findViewById(R.id.tv_submit);
    mBlack.setOnClickListener(this);
    mSubmint.setOnClickListener(this);
}

private void initData() {
    mtitleTextView.setText("费用申请");
}

/***
 * 提交数据之前进行一些判断,减轻服务器压力
 */
private void submit() {
    String money = mMoney.getText().toString().trim();
    float floatMoney = Float.parseFloat(money);
    if (floatMoney > 100000) {
        Toast.makeText(CostApplyActivity.this, "申请金额不能超过十万",
                Toast.LENGTH_LONG).show();
        return;
    }
    String reason = mReason.getText().toString().trim();
    if (money.equals("") || money == null) {
        Toast.makeText(CostApplyActivity.this, "请输入申请的金额",
                Toast.LENGTH_LONG).show();
        return;
    }
    if (reason.equals("") || reason == null) {
        Toast.makeText(CostApplyActivity.this, "请输入申请的理由",
                Toast.LENGTH_LONG).show();
        return;
    }

    Toast.makeText(CostApplyActivity.this, "提交成功,等待处理",
        Toast.LENGTH_LONG).show();
```

```
        finish();

    }

    @Override
    public void onClick(View v) {
        switch (v.getId()) {
        case R.id.iv_black:// 返回
            this.finish();
            break;
        case R.id.tv_submit:// 提交
            submit();
            break;
        }
    }
}
```

9.7 请假模块

此模块包括两个页面：请假列表以及请假申请，功能是展示请假信息、提交请假申请。请假列表主要包含的信息包括请假列表的标题，请假的条目信息展示(包括请假人、请假类别、开始时间和结束时间)，效果如图 9-13 所示；请假申请的信息包括请假申请的标题、返回和提交按钮、开始时间和结束时间的选择、请假类别的选择以及填写请假的理由，效果如图 9-13 所示。

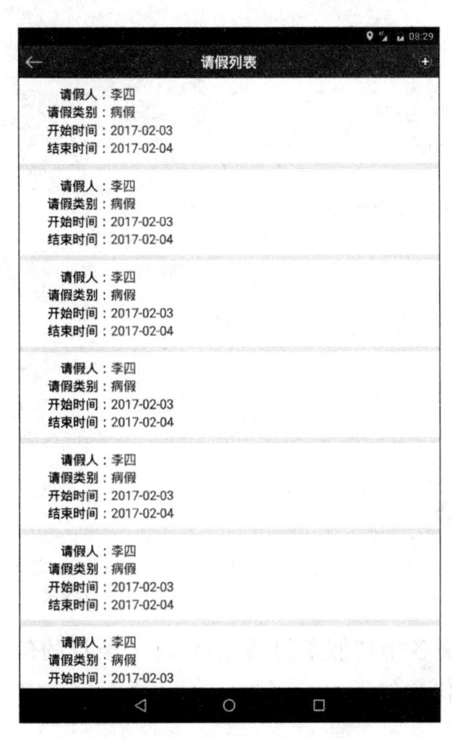

图 9-13 请假列表效果

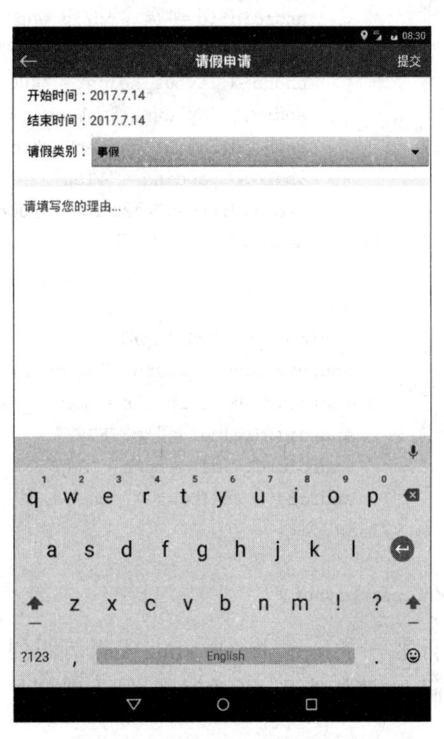

图 9-14 请假申请效果

9.7.1 请假列表

在 res/drawable-hdpi 中用到的资源文件如表 9-7 所示。

表 9-7 图片资源

图片名称	图　片	说　明
add_leave		为在 Word 上易于辨识，此处为该图加上了背景色

在 res/layout 下新建一个 activity_leave.xml 文件，实现请假列表总体显示，其中用<include/>标签作为布局引用。主要代码如下：

```
<?xml version = "1.0" encoding = "utf - 8"?>
<LinearLayout xmlns:android = "http://schemas.android.com/apk/res/android"
    android:layout_width = "match_parent"
    android:layout_height = "match_parent"
    android:orientation = "vertical" >

    <RelativeLayout
        android:layout_width = "match_parent"
        android:layout_height = "wrap_content" >

        <include layout = "@layout/view_title" />

        <ImageView
            android:id = "@ + id/iv_add"
            android:layout_width = "40dp"
            android:layout_height = "40dp"
            android:layout_alignParentRight = "true"
            android:layout_centerVertical = "true"
            android:padding = "11dp"
            android:src = "@drawable/add_leave" />
    </RelativeLayout>

    <ListView
        android:id = "@ + id/lstv"
        android:layout_width = "match_parent"
        android:layout_height = "match_parent"
        android:divider = "#F7F7F7"
        android:dividerHeight = "8dp"
        android:scrollbars = "none" >
    </ListView>

</LinearLayout>
```

新建一个 item_leave.xml，实现请假列表页面各个请假条目的显示，每个条目的信息包括请假人、请假类别、开始时间、结束时间。主要代码如下：

```
<?xml version = "1.0" encoding = "utf - 8"?>
```

```
<LinearLayout xmlns:android="http://schemas.android.com/apk/res/android"
    android:layout_width="match_parent"
    android:layout_height="wrap_content"
    android:orientation="vertical"
    android:paddingBottom="10dp"
    android:paddingLeft="20dp"
    android:paddingTop="10dp"
    android:background="#fff">

    <LinearLayout
        android:layout_width="match_parent"
        android:layout_height="wrap_content"
        android:orientation="horizontal">

        <TextView
            android:layout_width="110dp"
            android:layout_height="wrap_content"
            android:ellipsize="end"
            android:gravity="right|center_vertical"
            android:singleLine="true"
            android:text="请假人："
            android:textSize="18sp" />

        <TextView
            android:id="@+id/tv_person"
            android:layout_width="match_parent"
            android:layout_height="wrap_content"
            android:ellipsize="end"
            android:singleLine="true"
            android:textColor="@color/bg_title"
            android:textSize="18sp" />
    </LinearLayout>

    <LinearLayout
        android:layout_width="match_parent"
        android:layout_height="wrap_content"
        android:orientation="horizontal">

        <TextView
            android:layout_width="110dp"
            android:layout_height="wrap_content"
            android:ellipsize="end"
            android:gravity="right|center_vertical"
            android:singleLine="true"
            android:text="请假类别："
            android:textSize="18sp" />

        <TextView
            android:id="@+id/tv_type"
            android:layout_width="match_parent"
            android:layout_height="wrap_content"
```

```
            android:ellipsize = "end"
            android:singleLine = "true"
            android:textColor = "@color/bg_title"
            android:textSize = "18sp" />
</LinearLayout>

<LinearLayout
        android:layout_width = "match_parent"
        android:layout_height = "wrap_content"
        android:orientation = "horizontal" >

        <TextView
            android:layout_width = "110dp"
            android:layout_height = "wrap_content"
            android:ellipsize = "end"
            android:gravity = "right|center_vertical"
            android:singleLine = "true"
            android:text = "开始时间: "
            android:textSize = "18sp" />

        <TextView
            android:id = "@ + id/tv_start_time"
            android:layout_width = "match_parent"
            android:layout_height = "wrap_content"
            android:ellipsize = "end"
            android:singleLine = "true"
            android:textColor = "@color/bg_title"
            android:textSize = "18sp" />
</LinearLayout>

<LinearLayout
        android:layout_width = "match_parent"
        android:layout_height = "wrap_content"
        android:orientation = "horizontal" >

        <TextView
            android:layout_width = "110dp"
            android:layout_height = "wrap_content"
            android:ellipsize = "end"
            android:gravity = "right|center_vertical"
            android:singleLine = "true"
            android:text = "结束时间: "
            android:textSize = "18sp" />

        <TextView
            android:id = "@ + id/tv_end_time"
            android:layout_width = "match_parent"
            android:layout_height = "wrap_content"
            android:ellipsize = "end"
            android:singleLine = "true"
            android:textColor = "@color/bg_title"
```

```
            android:textSize = "18sp" />
    </LinearLayout >

</LinearLayout >
```

新建一个 LeaveInfo. java 实体类并进行封装，主要代码如下：

```
package com.qsd.kqxt.bean;

public class LeaveInfo {

    private int id;

    private String beginDate;

    private String endDate;

    private String leaveStaffName;

    private String typeName;

    public int getId() {
        return id;
    }

    public void setId(int id) {
        this.id = id;
    }

    public String getBeginDate() {
        return beginDate;
    }

    public void setBeginDate(String beginDate) {
        this.beginDate = beginDate;
    }

    public String getEndDate() {
        return endDate;
    }

    public void setEndDate(String endDate) {
        this.endDate = endDate;
    }

    public String getLeaveStaffName() {
        return leaveStaffName;
    }

    public void setLeaveStaffName(String leaveStaffName) {
        this.leaveStaffName = leaveStaffName;
```

```
    }

    public String getTypeName() {
        return typeName;
    }

    public void setTypeName(String typeName) {
        this.typeName = typeName;
    }

}
```

在 com. qsd. kqxt. ctl 目录下新建 LeaveAdapter. java 文件，该类继承 BaseAdapter 并实现了其中的方法，实现了 item_leave. xml 的列表显示。主要代码如下：

```
package com.qsd.kqxt.ctl;

import java.util.List;

import android.content.Context;
import android.view.View;
import android.view.ViewGroup;
import android.widget.BaseAdapter;
import android.widget.TextView;

import com.qsd.kqxt.R;
import com.qsd.kqxt.bean.LeaveInfo;

public class LeaveAdapter extends BaseAdapter {

    List<LeaveInfo> infos;
    Context context;

    public LeaveAdapter(List<LeaveInfo> infos, Context context) {
        this.infos = infos;
        this.context = context;
    }

    @Override
    public int getCount() {
        return infos.size();
    }

    @Override
    public Object getItem(int position) {
        return null;
    }

    @Override
    public long getItemId(int position) {
        return 0;
```

```
    }

    @Override
    public View getView(int position, View convertView, ViewGroup parent) {
        ViewHolder viewHolder;
        View view;
        if (convertView == null) {
            view = View.inflate(context, R.layout.item_leave, null);
            viewHolder = new ViewHolder();
            viewHolder.mName = (TextView) view.findViewById(R.id.tv_person);
            viewHolder.mType = (TextView) view.findViewById(R.id.tv_type);
            viewHolder.mStartTime = (TextView) view
                    .findViewById(R.id.tv_start_time);
            viewHolder.mEndTime = (TextView) view
                    .findViewById(R.id.tv_end_time);
            view.setTag(viewHolder);
        } else {
            view = convertView;
            viewHolder = (ViewHolder) convertView.getTag();
        }
        viewHolder.mName.setText(infos.get(position).getLeaveStaffName());
        viewHolder.mType.setText(infos.get(position).getTypeName());
        viewHolder.mStartTime.setText(infos.get(position).getBeginDate());
        viewHolder.mEndTime.setText(infos.get(position).getEndDate());

        return view;
    }

    public class ViewHolder {
        TextView mName;
        TextView mType;
        TextView mStartTime;
        TextView mEndTime;
    }

}
```

在 com.qsd.kqxt.ui 目录下新建一个 LeaveActivity.java，实现请假列表头部标题信息的设置，调用 LeaveAdapter 实现多条 item 的显示，代码如下：

```
package com.qsd.kqxt.ui;

import java.util.ArrayList;
import java.util.List;

import android.app.Activity;
import android.content.Intent;
import android.os.Bundle;
import android.view.View;
import android.view.View.OnClickListener;
import android.widget.AdapterView;
```

```
import android.widget.AdapterView.OnItemClickListener;
import android.widget.ImageView;
import android.widget.ListView;
import android.widget.TextView;

import com.qsd.kqxt.R;
import com.qsd.kqxt.adapter.LeaveAdapter;
import com.qsd.kqxt.bean.LeaveInfo;

public class LeaveActivity extends Activity implements OnClickListener {

    private TextView mtitleTextView;
    private ListView lstv;
    private ImageView mBlack, mAdd;
    int mState;

    int page = 1;

    List<LeaveInfo> infos = new ArrayList<LeaveInfo>();
    LeaveAdapter adapter;


    @Override
    protected void onCreate(Bundle savedInstanceState) {
        super.onCreate(savedInstanceState);
        setContentView(R.layout.activity_leave);

        initView();
        initData();
    }

    private void initData() {
        for (int i = 0; i < 16; i++) {
            LeaveInfo info = new LeaveInfo();
            info.setId(i);
            info.setLeaveStaffName("李四");
            info.setBeginDate("2017-02-03");
            info.setEndDate("2017-02-04");
            info.setTypeName("病假");
            infos.add(info);
        }
        adapter.notifyDataSetChanged();
    }

    private void initView() {

        mtitleTextView = (TextView) findViewById(R.id.tv_title);
        lstv = (ListView) findViewById(R.id.lstv);
        mBlack = (ImageView) findViewById(R.id.iv_black);
        mAdd = (ImageView) findViewById(R.id.iv_add);
```

```
        mBlack.setOnClickListener(this);
        mAdd.setOnClickListener(this);

        mtitleTextView.setText("请假列表");
        adapter = new LeaveAdapter(infos, LeaveActivity.this);
        lstv.setAdapter(adapter);
        lstv.setOnItemClickListener(new OnItemClickListener() {

            @Override
            public void onItemClick(AdapterView<?> arg0, View arg1, int arg2,
                    long arg3) {
                Intent intent = new Intent(LeaveActivity.this,
                        LeaveDetilActivity.class);
                intent.putExtra("leaveid", infos.get(arg2).getId() + "");
                startActivity(intent);
            }
        });
    }

    @Override
    public void onClick(View v) {
        // TODO Auto-generated method stub
        switch (v.getId()) {
        case R.id.iv_black:// 返回
            this.finish();
            break;
        case R.id.iv_add:// 请假申请
            Intent intent = new Intent(LeaveActivity.this,
                    LeaveApplyActivity.class);
            startActivity(intent);
            break;
        default:
            break;
        }
    }

}
```

9.7.2　请假申请

在 res/layout 下新建一个布局文件 activity_leave_apply.xml，实现请假申请页面的显示。主要代码如下：

```
<?xml version = "1.0" encoding = "utf-8"?>
<LinearLayout xmlns:android = "http://schemas.android.com/apk/res/android"
    android:layout_width = "match_parent"
    android:layout_height = "match_parent"
    android:background = "#fff"
    android:orientation = "vertical" >
```

```
<RelativeLayout
    android:layout_width="match_parent"
    android:layout_height="wrap_content" >

    <include layout="@layout/view_title" />

    <TextView
        android:id="@+id/tv_submit"
        android:layout_width="wrap_content"
        android:layout_height="45dp"
        android:layout_alignParentRight="true"
        android:layout_centerVertical="true"
        android:gravity="center"
        android:paddingLeft="5dp"
        android:paddingRight="15dp"
        android:text="提交"
        android:textColor="#fff"
        android:textSize="18sp" />
</RelativeLayout>

<View
    android:layout_width="match_parent"
    android:layout_height="2dp"
    android:background="#F9F9F9" />

<LinearLayout
    android:layout_width="match_parent"
    android:layout_height="wrap_content"
    android:layout_marginTop="10dp"
    android:orientation="horizontal" >

    <TextView
        android:layout_width="110dp"
        android:layout_height="wrap_content"
        android:ellipsize="end"
        android:gravity="right|center_vertical"
        android:singleLine="true"
        android:text="开始时间:"
        android:textSize="18sp" />

    <TextView
        android:id="@+id/tv_start_time"
        android:layout_width="match_parent"
        android:layout_height="wrap_content"
        android:ellipsize="end"
        android:singleLine="true"
        android:textColor="@color/bg_title"
        android:textSize="18sp" />
</LinearLayout>

<LinearLayout
```

```
    android:layout_width = "match_parent"
    android:layout_height = "wrap_content"
    android:layout_marginTop = "10dp"
    android:orientation = "horizontal" >

    <TextView
        android:layout_width = "110dp"
        android:layout_height = "wrap_content"
        android:ellipsize = "end"
        android:gravity = "right|center_vertical"
        android:singleLine = "true"
        android:text = "结束时间: "
        android:textSize = "18sp" />

    <TextView
        android:id = "@ + id/tv_end_time"
        android:layout_width = "match_parent"
        android:layout_height = "wrap_content"
        android:ellipsize = "end"
        android:singleLine = "true"
        android:textColor = "@color/bg_title"
        android:textSize = "18sp" />
</LinearLayout>

<LinearLayout
    android:layout_width = "match_parent"
    android:layout_height = "wrap_content"
    android:layout_marginBottom = "10dp"
    android:layout_marginTop = "10dp"
    android:orientation = "horizontal" >

    <TextView
        android:layout_width = "110dp"
        android:layout_height = "wrap_content"
        android:ellipsize = "end"
        android:gravity = "right|center_vertical"
        android:singleLine = "true"
        android:text = "请假类别: "
        android:textSize = "18sp" />

    <Spinner
        android:id = "@ + id/spinner"
        android:layout_width = "match_parent"
        android:layout_height = "wrap_content" />
</LinearLayout>

<View
    android:layout_width = "match_parent"
    android:layout_height = "10dp"
    android:background = "#F9F9F9" />
```

```
    <EditText
        android:id="@+id/et_reason"
        android:layout_width="match_parent"
        android:layout_height="0dp"
        android:layout_weight="1"
        android:background="@null"
        android:gravity="top"
        android:hint="请填写您的理由..."
        android:padding="15dp" />

</LinearLayout>
```

在 com.qsd.kqxt.ui 包下新建一个类 LeaveApplyActivity.java，主要代码如下：

```
package com.qsd.kqxt.ui;

import java.text.ParseException;
import java.text.SimpleDateFormat;
import java.util.ArrayList;
import java.util.Calendar;
import java.util.Date;
import java.util.HashMap;
import java.util.List;
import java.util.Map;

import android.app.ActionBar.LayoutParams;
import android.app.Activity;
import android.app.AlertDialog;
import android.app.Dialog;
import android.os.Bundle;
import android.util.Log;
import android.view.LayoutInflater;
import android.view.View;
import android.view.View.OnClickListener;
import android.widget.AdapterView;
import android.widget.AdapterView.OnItemClickListener;
import android.widget.ArrayAdapter;
import android.widget.DatePicker;
import android.widget.ImageView;
import android.widget.ListView;
import android.widget.PopupWindow;
import android.widget.Spinner;
import android.widget.TextView;
import android.widget.Toast;

import com.qsd.kqxt.R;
import com.qsd.kqxt.adapter.TypesPopWindowAdapter;

public class LeaveApplyActivity extends Activity implements OnClickListener {

    private TextView mtitleTextView;
```

```
private TextView mStartTime;
private TextView mEndTime;
private TextView mType;
private TextView mReason;
private ImageView mBlack;
private TextView mSubmit;

private Spinner spinner;
private List<String> types;
private ArrayAdapter<String> arr_adapter;
PopupWindow popupWindow;
Dialog dialog;
DatePicker mDP;

int timeState;
int type = -1;

int year;
int month;
int day;

@Override
protected void onCreate(Bundle savedInstancetimeState) {
    super.onCreate(savedInstancetimeState);
    setContentView(R.layout.activity_leave_apply);

    initView();
    initTime();
    initData();
    setDialog();
}

private void initView() {
    mtitleTextView = (TextView) findViewById(R.id.tv_title);
    mStartTime = (TextView) findViewById(R.id.tv_start_time);
    spinner = (Spinner) findViewById(R.id.spinner);
    mEndTime = (TextView) findViewById(R.id.tv_end_time);
    mType = (TextView) findViewById(R.id.tv_type);
    mReason = (TextView) findViewById(R.id.et_reason);
    mSubmit = (TextView) findViewById(R.id.tv_submit);
    mBlack = (ImageView) findViewById(R.id.iv_black);
    mBlack.setOnClickListener(this);
    mSubmit.setOnClickListener(this);

}

private void initTime() {
    Calendar c = Calendar.getInstance();
    year = c.get(Calendar.YEAR);
    month = c.get(Calendar.MONTH) + 1;
    day = c.get(Calendar.DAY_OF_MONTH);
```

```
        String dayString = day + "";
        if (day < 10) {
            dayString = "0" + day;
        }

        mStartTime.setText(year + "." + month + "." + dayString);
        mEndTime.setText(year + "." + month + "." + dayString);
    }

    private void setDialogListener() {
        mStartTime.setOnClickListener(new OnClickListener() {
            @Override
            public void onClick(View v) {
                timeState = 0;
                dialog.show();
            }
        });
        mEndTime.setOnClickListener(new OnClickListener() {
            @Override
            public void onClick(View v) {
                timeState = 1;
                dialog.show();
            }
        });
    }

    private void setDialog() {
        View view = View.inflate(LeaveApplyActivity.this, R.layout.view_dialog,
                null);
        mDP = (DatePicker) view.findViewById(R.id.data_picker);
        TextView mOk = (TextView) view.findViewById(R.id.tv_ok);
        mOk.setOnClickListener(new OnClickListener() {
            @Override
            public void onClick(View v) {
                year = mDP.getYear();
                month = mDP.getMonth() + 1;
                day = mDP.getDayOfMonth();
                String dayString = day + "";
                if (day < 10) {
                    dayString = "0" + day;
                }
                if (timeState == 0) {
                    mStartTime.setText(year + "." + month + "." + dayString);
                } else {
                    mEndTime.setText(year + "." + month + "." + dayString);
                }
                dialog.dismiss();
            }
        });
```

```
        dialog = new AlertDialog.Builder(LeaveApplyActivity.this).create();
        dialog.show();
        dialog.setContentView(view);
        dialog.dismiss();
        setDialogListener();
    }

    private void initData() {

        mtitleTextView.setText("请假申请");
        types = new ArrayList<String>();
        types.add("事假");
        types.add("病假");
        types.add("婚假");
        types.add("产假");
        arr_adapter = new ArrayAdapter<String>(this,
                android.R.layout.simple_spinner_item, types);
        arr_adapter
                .setDropDownViewResource(android.R.layout.simple_spinner_dropdown_item);
        spinner.setAdapter(arr_adapter);

    }

    /***
     * 判断时间的先后
     * @return
     */
    private boolean compareTime() {
        String startTime = mStartTime.getText().toString().trim();
        String endTime = mEndTime.getText().toString().trim();
        SimpleDateFormat format = new SimpleDateFormat("yyyy.MM.dd");
        try {
            Date startData = format.parse(startTime);
            Date endData = format.parse(endTime);
            if (startData.getTime() <= endData.getTime()) {
                return false;
            } else {
                return true;
            }
        } catch (ParseException e1) {
            e1.printStackTrace();
        }
        return false;
    }

    @Override
    public void onClick(View v) {
        // TODO Auto-generated method stub
```

```
            switch (v.getId()) {
            case R.id.iv_black:// 返回
                this.finish();
                break;
            case R.id.tv_submit:// 提交
                if (compareTime()) {
                    Toast.makeText(this, "结束时间不能小于开始时间", 1).show();
                } else {
                    finish();
                }
                break;
            }
        }
    }
```

9.8 设置模块

此模块主要功能是修改密码和用户退出。包含的页面有设置页面、修改密码页面。从设计效果图 9-15 看,设置页面主要包括头部标题、后退按钮、修改密码和用户退出按钮。从设计效果图 9-16 看,修改密码页面主要包括头部信息、提交和后退按钮,原始密码、新密码、确认密码的输入文本域。退出确认的效果如图 9-17 所示。

图 9-15 设置页面效果

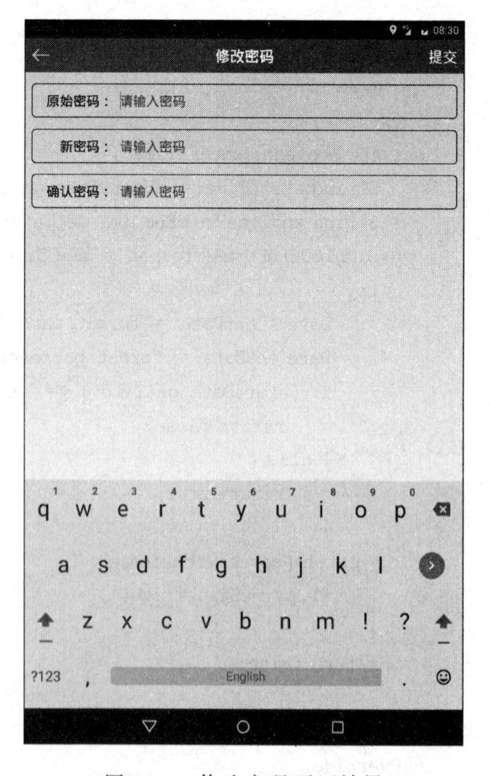

图 9-16 修改密码页面效果

图 9-17　退出确认

9.8.1　修改密码

在 res/layout 下添加 activity_change_pwd.xml，实现修改密码页面的头部标题引入（include 标签）、提交按钮、输入原始密码、新密码、确认密码区域的显示。主要代码如下：

```
<?xml version = "1.0" encoding = "utf - 8"?>
<LinearLayout xmlns:android = "http://schemas.android.com/apk/res/android"
    android:layout_width = "match_parent"
    android:layout_height = "match_parent"
    android:orientation = "vertical" >

    <RelativeLayout
        android:layout_width = "match_parent"
        android:layout_height = "wrap_content" >

        <include layout = "@layout/view_title" />

        <TextView
            android:id = "@ + id/tv_submit"
            android:layout_width = "wrap_content"
            android:layout_height = "wrap_content"
            android:layout_alignParentRight = "true"
            android:layout_centerVertical = "true"
            android:gravity = "center_vertical"
```

```
            android:paddingRight = "10dp"
            android:text = "提交"
            android:textColor = "#fff"
            android:textSize = "20sp" />
</RelativeLayout>

<LinearLayout
        android:layout_width = "match_parent"
        android:layout_height = "45dp"
        android:layout_marginLeft = "10dp"
        android:layout_marginRight = "10dp"
        android:layout_marginTop = "15dp"
        android:background = "@drawable/shape_edit_bg"
        android:orientation = "horizontal" >

    <TextView
            android:layout_width = "110dp"
            android:layout_height = "match_parent"
            android:gravity = "center_vertical|right"
            android:text = "原始密码: "
            android:textSize = "18sp" />

    <EditText
            android:id = "@ + id/old_password"
            android:layout_width = "match_parent"
            android:layout_height = "match_parent"
            android:background = "@null"
            android:gravity = "center_vertical"
            android:hint = "请输入密码"
            android:inputType = "textPassword"
            android:paddingLeft = "10dp" />
</LinearLayout>

<LinearLayout
        android:layout_width = "match_parent"
        android:layout_height = "45dp"
        android:layout_marginLeft = "10dp"
        android:layout_marginRight = "10dp"
        android:layout_marginTop = "15dp"
        android:background = "@drawable/shape_edit_bg"
        android:orientation = "horizontal" >

    <TextView
            android:layout_width = "110dp"
            android:layout_height = "match_parent"
            android:gravity = "center_vertical|right"
            android:text = "新密码: "
            android:textSize = "18sp" />

    <EditText
            android:id = "@ + id/new_password"
```

```
            android:layout_width = "match_parent"
            android:layout_height = "match_parent"
            android:background = "@null"
            android:gravity = "center_vertical"
            android:hint = "请输入密码"
            android:inputType = "textPassword"
            android:paddingLeft = "10dp" />
    </LinearLayout>

    <LinearLayout
        android:layout_width = "match_parent"
        android:layout_height = "45dp"
        android:layout_marginLeft = "10dp"
        android:layout_marginRight = "10dp"
        android:layout_marginTop = "15dp"
        android:background = "@drawable/shape_edit_bg"
        android:orientation = "horizontal" >

        <TextView
            android:layout_width = "110dp"
            android:layout_height = "match_parent"
            android:gravity = "center_vertical|right"
            android:text = "确认密码: "
            android:textSize = "18sp" />

        <EditText
            android:id = "@ + id/new_password2"
            android:layout_width = "match_parent"
            android:layout_height = "match_parent"
            android:background = "@null"
            android:gravity = "center_vertical"
            android:hint = "请输入密码"
            android:inputType = "textPassword"
            android:paddingLeft = "10dp" />
    </LinearLayout>

</LinearLayout>
```

在 com.qsd.kqxt.activity 包下添加 ChangePwdActivity 类。该类实现了修改密码页面的头部标题信息的设置，对用户提交的原始密码、新密码、确认密码的后台校验以及信息提交功能。主要代码如下：

```
package com.qsd.kqxt.activity;

import com.qsd.kqxt.R;

import android.app.Activity;
import android.os.Bundle;
import android.view.View;
import android.view.View.OnClickListener;
import android.widget.EditText;
```

```
import android.widget.ImageView;
import android.widget.TextView;
import android.widget.Toast;

public class ChangePwdActivity extends Activity {

    EditText oldPwd;
    EditText newPwd;
    EditText newPwd2;
    TextView submit;

    ImageView back;
    TextView title;

    @Override
    protected void onCreate(Bundle savedInstanceState) {
        super.onCreate(savedInstanceState);
        setContentView(R.layout.activity_change_pwd);
        initView();
        setListener();
    }

    private void initView() {
        oldPwd = (EditText) findViewById(R.id.old_password);
        newPwd = (EditText) findViewById(R.id.new_password);
        newPwd2 = (EditText) findViewById(R.id.new_password2);
        submit = (TextView) findViewById(R.id.tv_submit);
        back = (ImageView) findViewById(R.id.iv_black);
        title = (TextView) findViewById(R.id.tv_title);
        title.setText("修改密码");
    }

    private void setListener() {
        back.setOnClickListener(new OnClickListener() {
            @Override
            public void onClick(View v) {
                finish();
            }
        });
        submit.setOnClickListener(new OnClickListener() {
            @Override
            public void onClick(View v) {
                if (checkPwd()) {
                    submit();
                }
            }
        });
    }

    private boolean checkPwd() {
        String oldPwdString = oldPwd.getText().toString().trim();
```

```
        String newPwdString1 = newPwd.getText().toString().trim();
        String newPwdString2 = newPwd2.getText().toString().trim();
        if (oldPwdString == null || oldPwdString.equals("")) {
            Toast.makeText(ChangePwdActivity.this, "请输入原始密码",
                    Toast.LENGTH_LONG).show();
            return false;
        }
        if (newPwdString1 == null || newPwdString1.equals("")) {
            Toast.makeText(ChangePwdActivity.this, "请输入新密码",
                Toast.LENGTH_LONG)
                    .show();
            return false;
        }
        if (!newPwdString1.equals(newPwdString2)) {
            Toast.makeText(ChangePwdActivity.this, "两次输入的密码不一致",
                    Toast.LENGTH_LONG).show();
            return false;
        }
        if (newPwdString1.length() < 6) {
            Toast.makeText(ChangePwdActivity.this, "密码必须大于 6 位",
                    Toast.LENGTH_LONG).show();
            return false;
        }
        return true;
    }

    private void submit() {
        Toast.makeText(ChangePwdActivity.this, "提交成功", 2000).show();
    }
}
```

9.8.2 用户退出

在 res/layout 文件下新建 activity_setting.xml,实现设置页面的显示,包括头部信息的引用(include 标签)、修改密码和用户退出的区域的显示。主要代码如下:

```
<?xml version = "1.0" encoding = "utf - 8"?>
<LinearLayout xmlns:android = "http://schemas.android.com/apk/res/android"
    android:layout_width = "match_parent"
    android:layout_height = "match_parent"
    android:orientation = "vertical" >

    <include layout = "@layout/view_title" />

    <LinearLayout
        android:layout_width = "match_parent"
        android:layout_height = "wrap_content"
        android:layout_marginLeft = "15dp"
        android:layout_marginRight = "15dp"
        android:layout_marginTop = "15dp"
```

```
        android:background = "@drawable/shape_edit_bg"
        android:orientation = "vertical" >

        <TextView
            android:id = "@ + id/tv_change_pwd"
            android:layout_width = "match_parent"
            android:layout_height = "wrap_content"
            android:padding = "10dp"
            android:text = "修改密码"
            android:textSize = "18sp" />

        <View
            android:layout_width = "match_parent"
            android:layout_height = "1dp"
            android:background = "#000" />

        <TextView
            android:id = "@ + id/tv_exit"
            android:layout_width = "match_parent"
            android:layout_height = "wrap_content"
            android:padding = "10dp"
            android:text = "用户退出"
            android:textSize = "20sp" />
    </LinearLayout>

</LinearLayout>
```

在 com. qsd. kqxt. bean 包下添加 UserInfo. java，该类将在 SettingActivity. java 中被调用。主要代码如下：

```
package com.qsd.kqxt.bean;

public class UserInfo {

    public static String adminID;
    public static String adminName;
    public static String adminPwd;
    public static String staffName;
    public static String staffTel;
    public static String departID;
    public static String departName;
    public static String roleID;
    public static String roleName;

}
```

在 com. qsd. kqxt. activity 文件下新建 SettingActivity. java。该类实现对设置页面的标题信息的设置、调用 ChangePwdActivity 实现密码的修改以及用户退出的功能。主要代码如下：

```
package com.qsd.kqxt.activity;
```

```
import android.app.Activity;
import android.app.AlertDialog;
import android.content.DialogInterface;
import android.content.Intent;
import android.os.Bundle;
import android.view.View;
import android.view.View.OnClickListener;
import android.widget.ImageView;
import android.widget.TextView;

import com.qsd.kqxt.R;
import com.qsd.kqxt.LoginActivity;
import com.qsd.kqxt.MainActivity;
import com.qsd.kqxt.bean.UserInfo;
import com.qsd.kqxt.utils.SPUtils;

public class SettingActivity extends Activity {

    TextView pwd;
    TextView exit;
    ImageView back;
    TextView title;

    AlertDialog dialog;

    SPUtils utils;

    @Override
    protected void onCreate(Bundle savedInstanceState) {
        super.onCreate(savedInstanceState);
        setContentView(R.layout.activity_setting);
        utils = new SPUtils(SettingActivity.this);
        initView();
        setListener();
        setDialog();
    }

    private void initView() {
        pwd = (TextView) findViewById(R.id.tv_change_pwd);
        exit = (TextView) findViewById(R.id.tv_exit);
        back = (ImageView) findViewById(R.id.iv_black);
        title = (TextView) findViewById(R.id.tv_title);
        title.setText("设置");
    }

    private void setListener() {
        pwd.setOnClickListener(new OnClickListener() {
            @Override
            public void onClick(View v) {
                toChangePwdActivity();
```

```
            }
        });
        exit.setOnClickListener(new OnClickListener() {
            @Override
            public void onClick(View v) {
                dialog.show();
            }
        });
        back.setOnClickListener(new OnClickListener() {
            @Override
            public void onClick(View v) {
                finish();
            }
        });
    }

    private void toChangePwdActivity() {
        Intent intent = new Intent(SettingActivity.this,
                ChangePwdActivity.class);
        startActivity(intent);
    }

    private void setDialog() {
        AlertDialog.Builder builder = new AlertDialog.Builder(
        SettingActivity.this, android.R.style.Theme_Holo_Light_Dialog);
        builder.setMessage("您确定要退出吗?");
        builder.setPositiveButton("确定", new DialogInterface
            .OnClickListener(){
            @Override
            public void onClick(DialogInterface dialog, int which) {
                exit();
            }
        });
        builder.setNegativeButton("取消", new DialogInterface
            .OnClickListener(){
            @Override
            public void onClick(DialogInterface dialog, int which) {
                dialog.dismiss();
            }
        });
        dialog = builder.create();
        dialog.getWindow().setBackgroundDrawableResource(
                android.R.color.transparent);
    }

    private void exit() {
        removeData();
        Intent intent = new Intent(SettingActivity.this, LoginActivity.class);
        startActivity(intent);
        MainActivity.instance.finish();
        finish();
```

```
    }

    private void removeData() {
        utils.clearAllData();
        UserInfo.adminID = null;
        UserInfo.adminID = null;
        UserInfo.adminName = null;
        UserInfo.departID = null;
        UserInfo.departName = null;
        UserInfo.roleID = null;
        UserInfo.roleName = null;
        UserInfo.staffName = null;
        UserInfo.staffTel = null;
    }
}
```

参 考 文 献

[1] 智能手机操作系统[EB/OL].(2016-04-01)[2017-09-04]
https://zhidao.baidu.com/question/2205234664126639748.html.
[2] 青岛东合信息技术有限公司. Android 程序设计[M]. 北京:电子工业出版社,2012.
[3] 传智播客高教产品研发部. Android 移动应用基础教程[M]. 北京:中国铁道出版社,2015.
[4] Jackson W. Android 应用开发入门[M]. 周自恒,译. 北京:人民邮电出版社,2013.
[5] Android 数据存储[EB/OL].(2013-09-13)[2017-08-12]
https://my.oschina.net/handsomeban/blog/207025.
[6] SQLite 数据库[EB/OL].(2013-04-13)[2017-08-15]
http://www.cnblogs.com/menlsh/archive/2013/04/13/3019588.html.
[7] 使用 Content provider 方式共享数据[EB/OL].(2013-04-17)[2017-08-17]
http://www.cnblogs.com/menlsh/archive/2013/04/17/3027394.html.
[8] 网络编程[EB/OL].(2016-02-19)[2017-08-20]
https://wenku.baidu.com/view/ca89a99d69eae009591bec65.html.
[9] 李刚. 疯狂 Android 讲义[M]. 3 版. 北京:电子工业出版社,2015.
[10] Intent 与组件通信[EB/OL].(2014-09-06)[2017-08-22]
http://www.cnblogs.com/smyhvae/p/3959204.html.
[11] Android 官方文档之 Service[EB/OL].(2016-05-11)[2017-08-23]
http://blog.csdn.net/vanpersie_9987/article/details/51360245.
[12] 浅谈 Android 中的 Service[EB/OL].(2016-07-24)[2017-08-23]
http://blog.csdn.net/u012318003/article/details/52015853.
[13] Android 四大组件之 Service[EB/OL].(2016-03-01)[2017-08-24]
http://blog.csdn.net/seebetpro/article/details/50195077.
[14] 关于闹钟设置 AlarmManager 类方法参数解释[EB/OL].(2014-07-04)[2017-08-25]
http://www.cnblogs.com/crazywenza/p/3823774.html.
[15] 郭霖. 第一行代码 Android[M]. 2 版. 北京:人民邮电出版社,2016.
[16] Android 定位 SDK[EB/OL].(2016-07-19)[2017-08-20]
http://lbsyun.baidu.com/index.php?title=android-locsdk/guide/getloc&qq-pf-to=pcqq.c2c.
[17] 包建强. App 研发录:架构设计、Crash 分析和竞品技术分析[M]. 北京:机械工业出版社,2015.

图书资源支持

感谢您一直以来对清华版图书的支持和爱护。为了配合本书的使用，本书提供配套的资源，有需求的读者请扫描下方的“书圈”微信公众号二维码，在图书专区下载，也可以拨打电话或发送电子邮件咨询。

如果您在使用本书的过程中遇到了什么问题，或者有相关图书出版计划，也请您发邮件告诉我们，以便我们更好地为您服务。

我们的联系方式：

地　　址：北京市海淀区双清路学研大厦 A 座 714

邮　　编：100084

电　　话：010-83470236　010-83470237

客服邮箱：2301891038@qq.com

QQ：2301891038（请写明您的单位和姓名）

资源下载：关注公众号“书圈”下载配套资源。

资源下载、样书申请

书圈

获取最新书目

观看课程直播

质检